# Hypnosis Between Science and Magic

*LINES*

**Series editors: Matthew Fuller and Andrew Goffey**

*Lines of flight, lines of code, lines of text, lines of thought, lines of paint, lines of powder, lines of conflict, lines of alliance, lines of connection...*

*Lines* is a series of intensely written books linking the broad field of contemporary theory to the large-scale phenomena of today.

Books in this series engage work which emerges from post-disciplinary interstices of critical and speculative thinking: treating conceptual mobility, disciplinary undecidability and aesthetic and formal imagination as critical and investigative strengths.

**Also Available from Bloomsbury**

*Philosophy and Psychedelics*, Christine Hauskeller and Peter Sjöstedt-Hughes
*Materialist Phenomenology*, Manuel DeLanda
*Technic and Magic*, Federico Campagna
*How to Sleep: The Art, Biology and Culture of Unconsciousness*, Matthew Fuller

# Hypnosis Between Science and Magic

*Isabelle Stengers*

Translated by April A. Knutson
Edited by Andrew Goffey

BLOOMSBURY ACADEMIC
LONDON · NEW YORK · OXFORD · NEW DELHI · SYDNEY

BLOOMSBURY ACADEMIC
Bloomsbury Publishing Plc
50 Bedford Square, London, WC1B 3DP, UK
1385 Broadway, New York, NY 10018, USA
29 Earlsfort Terrace, Dublin 2, Ireland

BLOOMSBURY, BLOOMSBURY ACADEMIC and the Diana logo are trademarks of
Bloomsbury Publishing Plc

First published in 2002 in France *as L'Hypnose entre magie et science* by EMPÊCHEURS De PENSER EN ROND

First published in Great Britain 2025

English language translation © April A. Knutson, 2025

April A. Knutson has asserted her right under the Copyright, Designs and Patents Act, 1988, to be identified as Translator of this work.

All rights reserved. No part of this publication may be reproduced or transmitted in any form or by any means, electronic or mechanical, including photocopying, recording, or any information storage or retrieval system, without prior permission in writing from the publishers.

Bloomsbury Publishing Plc does not have any control over, or responsibility for, any third-party websites referred to or in this book. All internet addresses given in this book were correct at the time of going to press. The author and publisher regret any inconvenience caused if addresses have changed or sites have ceased to exist, but can accept no responsibility for any such changes.

A catalogue record for this book is available from the British Library.

A catalog record for this book is available from the Library of Congress.
Names: Stengers, Isabelle, author. | Knutson, April, translator. | Goffey, Andrew, editor.
Title: Hypnosis between science and magic / Isabelle Stengers; translated by April A. Knutson; edited by Andrew Goffey.
Other titles: Hypnose entre magie et science. English
Description: London; New York: Bloomsbury Academic, 2024. | Series: Lines | "First published in 2002 in France as L'Hypnose entre magie et science by EMPÊCHEURS." | Includes bibliographical references and index.
Identifiers: LCCN 2024020728 (print) | LCCN 2024020729 (ebook) | ISBN 9781350501423 (hb) | ISBN 9781350501416 (pb) | ISBN 9781350501447 (ebook) | ISBN 9781350501430 (epdf)
Subjects: LCSH: Hypnotism–Therapeutic use. | Psychoanalysis.
Classification: LCC RC495 .S7413 2024 (print) | LCC RC495 (ebook) | DDC 615.8/512–dc23/eng/20240613
LC record available at https://lccn.loc.gov/2024020728
LC ebook record available at https://lccn.loc.gov/2024020729

ISBN: HB: 978-1-3505-0142-3
PB: 978-1-3505-0141-6
ePDF: 978-1-3505-0143-0
eBook: 978-1-3505-0144-7

Series: Lines

Typeset by Deanta Global Publishing Services, Chennai, India
Printed and bound in Great Britain

To find out more about our authors and books visit www.bloomsbury.com and sign up for our newsletters.

*To the memory of Léon Chertok, who learned how to think with hypnosis*

# Contents

Introduction: Magic Gestures: Stengers, Hypnosis and the Cultivation of Perplexity 1

1   Wounds 55

2   A History that Stutters 65

3   Lessons from History 81

4   Freud's Coup de Force 95

5   It's only an Artifact? 109

6   Thinking Therapeutic Techniques? 129

*Index* 155

# Introduction

## Magic Gestures: Stengers, Hypnosis, and the Cultivation of Perplexity

> Isn't it ridiculous that with respect to the phenomenon of hypnosis, whose enigmatic character Freud had always recognised, we are still at the level of invoking Hitler, drugs, or the music hall?[1]

*Hypnosis Between Science and Magic*, which was first published in France in 2002 as *L'hypnose entre magie et science*, is one of a number of books, articles, prefaces written by Isabelle Stengers on hypnosis. In Anglophone contexts, Stengers is generally known as a philosopher of science whose thinking is exemplified in her two-volume (originally seven) *Cosmopolitics* and her major study of philosopher Alfred North Whitehead *Thinking with Whitehead*, both published in English translation relatively recently.[2] But these better-known publications follow a number of other books by her, including those on hypnosis, that have not been given as much consideration, and they have even been somewhat overlooked in Anglophone engagements with her work.[3] Indeed, while a good deal of her writing since the *Cosmopolitics* essays has been translated, much of her quite considerable earlier work on hypnosis still exists in the original French only. Aside from the book *A Critique of Psychoanalytic Reason*, which she cowrote with Léon Chertok (of whom more later) and which was actually one of the first translations of her work into English[4] and the present book, Stengers also cowrote Chertok's (auto)biography, *Mémoires d'un hérétique*, first published in 1990,[5] a much shorter book, also with Chertok, on narcissistic wounds *L'hypnose: Blessure narcissique*,[6] has edited a collection on the importance of hypnosis,[7] coauthored a play (with Tobie Nathan and Lucien Hounkpatin) *La Damnation de Freud*,[8] as well as numerous interventions, prefaces, and articles.

For Stengers—and putting aside shifts and transitions in her thinking more generally—hypnosis seems to have been something that is good to think with from quite early on.[9] From her first book with the physical chemist Ilya Prigogine—with its allusions to "hypnons" and "sleepwalker" particles[10]— through repeated invocations of the 1784 episode of the Royal Commissions in France with their condemnation of Mesmerism, her characterization of Whitehead's God (in *Process and Reality*) in terms of hypnotic induction, *Virgin Mary and the Neutrino*'s consideration of experimental hypnosis and the speculative contrast that book proposes between idiots and sleepwalkers, hypnosis and its history raises troubling questions. In the context of intellectual habits predicated on unexamined or overstated ideas of progress and presumptions to judge in particular, this history offers an important resource for thinking creatively about science, non-science, and the ecology of practices more generally.

This introduction aims to situate *Hypnosis Between Science and Magic* within the broader trajectory of Stengers's writings, to provide readers of the present book who may be unfamiliar with her work with some points of connection to that now substantial body of writing. It's neither an "explanation" nor a substitute for the transformative experience that, it is hoped, an encounter with Stengers can provoke.

## An Enigma

Roughly one century separates the taking in hand of Anton Mesmer's animal magnetism by royally appointed scientific commissioners in prerevolutionary Paris from the positivist era that saw the emergence of hypnosis "proper." Associated, notably, with Freud's teacher Charcot at the Hopital Saltpetrière, who proposed a "somatic" account of hypnosis as the product of hysteria, as well as with figures such as Liébault and Bernheim (the "Nancy School"), for whom a purely psychological conception of suggestion took precedence, this era—effectively marking the birth of the "psy" disciplines more generally—saw the marginalization of the magnetists, against whom both Charcot and Bernheim, in their different ways, set themselves. Another century separates that period from the specific resurgence of interest in hypnosis in the last decades of the twentieth century with which Isabelle Stengers herself is connected.[11] This is not to say that nothing of interest had happened before or between these historical moments, of course, nor that Europe was the only region of the world in which mesmerism, magnetism, "trance," and healing practices predicated on "influence" were at play. Far from it. In the broader sweep of time, they form part of a proliferating—for some, worryingly uncontrollable—assortment of practices that continue to pose a challenge to modern assumptions about scientific expertise, the link between truth and health, the stable assigning of roles, especially in therapeutic situations, cherished notions such as freewill and so on. In Europe, before Mesmer and his famous "baquet" excited so much attention, it wasn't intellectual controversy or sneering derision that such practices excited so much as violent repression. Witchcraft, exorcism, faith healing (the Loudun possessions, the persecution of Jansenist priests, and the violent treatment of "convulsionaries" by means of "assistances,"[12] for example) are part of this broader historical fabric. So too are the widespread developments of magnetic somnambulism following the extraordinary discoveries of the Marquis de Puységur, after Mesmer, and the marvelous talents put into play by the "rapport" cultivated by magnetizers with their "subjects." The outflanking of the Marquis de Lafayette and the well-documented growth of spiritualism in the United States, debates about table-turning and clairvoyance, the Society for Psychical Research at Cambridge, Surrealist André Breton's "new consciousness," Cold War paranoia around brainwashing, complex relations to others (the history of Vaudou, for example) also belong to these developments. Hypnosis itself occupies an interesting position in this broader history, much of which has been repeatedly marginalized or forgotten.[13] Opposed to earlier practices of magnetism, for which it sought to offer more acceptably scientific forms of explanation, hypnosis too is marginalized following both the development and popularization of psychology and especially of psychoanalysis at the end of the nineteenth and start of the twentieth centuries. Experiencing something of a renaissance, in different ways, in the United States both before and after the Second World War, via the work of Clark Hall, Milton Erickson, Laurence Kubie, and others, the strange phenomena associated with hypnosis have been the subject of recurrent legal controversies, most notably concerning multiple personality disorders, and they continue to be subjected to the skeptical demands of experimental-psychological "proof," as has been the case since the 1920s.[14] Magnetists and hypnotists themselves might not have been burned at the stake but their marginalization has nonetheless been strikingly effective, even if for psychotherapy, and psychoanalysis in particular, this has come at the cost of what Isabelle Stengers's collaborator Léon Chertok could describe in 1998 as a major impasse and a situation of "quasi-total ignorance."[15]

For Stengers, addressing hypnosis now and taking seriously the questions that it raises obliges us to slow down, to hesitate, before rushing, armed with, say, the latest findings of neuroscience or the most sophisticated of theories, to dismiss it, explain its singularity away, or otherwise conclude that we now know it "really does exist" (or not). In this respect it challenges habitual judgments about what we "know" and others "believe," and the narratives of continuous, apparently ineluctable, progress that these judgments feed.[16] It can sensitize us to the stupid prerogatives of those that Stengers calls "our guardians," and the role of Science (with a capital "S" to indicate something that is rather different to the complex ecology of scientific and other practices) in the power that would take others (and it is always "others") in hand, on the pretext that letting them think for themselves would "open the floodgates" to all manner of demagogues, quacks, and charismatics.[17] It challenges the preeminent role that theory assumes in defining what it is that we are dealing with when it comes to "the" psyche, and taking it seriously necessitates a delicate engagement with history and the force it exercises on the kinds of questions that get asked about science, medicine, and knowledge practices more generally. In this respect the specific, heterogeneous ensemble of experiences that might broadly be qualified *as* hypnosis form, not so much a set of facts in search of a theory, but rather something that Stengers suggests "calls into question the judgement about 'reality' that a theory seeks to institute."[18] Needless to say, engaging with it also entails moving away from the rather clichéd images that have percolated throughout Western cultures, over the last, say 150 years, images that do very little to confer on hypnosis the capacity to make us think, and understanding something of how it has both disappointed and deceived those who sought to confer on it an explanatory power.

Well-sedimented distinctions between the human and the animal, conscious and unconscious "states," self and other, enlightened modern cultures and traditional practices, *fact* and *artefact* find themselves troubled by hypnosis. In *A Critique of Psychoanalytic Reason*, her first book on the topic (cowritten with Léon Chertok), Stengers referred to hypnosis as an "irreducibly biopsycho-sociological phenomenon," an "enigma"[19] that poses a challenge to the different ways in which scientific research has endeavored to engage with it, whether to stabilize it as an irrepressible trait of the human psyche, to explain (and so dismiss) it as nothing but compliant simulation or, indeed, to refer to it as evidence of the general credulity, passivity, and/or irrationality of human beings. Historically speaking, hypnosis has turned out not to be something one can scientifically define, and one will not find a definition of it in Stengers's work, at least not one that she subscribes to: definitions accord a stability to the phenomenon defined that in the case of hypnosis, knowledge practices have not successfully achieved for it. It's not even clear that we can talk about hypnosis as *one* thing. Mikkel Borch-Jacobsen, one of the relatively small number of philosophers to have taken it seriously in recent years, argues that

> hypnosis is but one name amongst many others for designating this elusive "X" that we try to approach. One can also call it "animal magnetism," "lucid sleep," or "transference" if one is more of a psychotherapist, "suggestion," "hysteria" or "modified state of consciousness" if one is more of a psychiatrist or psychologist, "trance" or "ecstasy" if one is an anthropologist, "demonic possession" if one is a theologian. And each one of these names will bring with it not only a different *theory* but also a different *phenomenon*, as if the most remarkable property of our "X" was to have none and to vary in concert with the discourse one holds about it.[20]

Deftly staged by Borch-Jacobsen,[21] the slipperiness of this "X" that could be called hypnosis (the name itself, as Stengers reminds us, derives from the Greek "hypnos" for "sleep," and it dates to 1843 and the work of James Braid, who originally referred to "neurypnology"), to say nothing of the one hundred-plus years of peripeteia around the findings of experimental psychological engagements with it, can lead to a simple conclusion, one noted right at the start of this book: that *it* does not exist.[22] Having deceived or disappointed the expectations of modern knowledge practices (and hypnosis hasn't stopped doing both of these),[23] one may feel, as a consequence, that it can simply be ignored, explained away, and thus that its instability need not oblige us especially to think.[24] Elsewhere in her work Stengers has noted that "our tradition" (that of the moderns) has a "preferred vice" which consists in "constructing a convenient argument that has, as if by accident, the power to dissimulate or silence a question it feels uncomfortable addressing."[25] It's a statement that is especially pertinent with regard to the questions hypnosis raises. Numerous strategies have exhibited this vice throughout the course of its "stuttering history," running the gamut from outright dismissal, through the use of catch-all terms that offer pseudo-explanations, to positions that rely on some or other version of the epic narrative of scientific discovery that draw our attention away from the impasses in human science research. Such strategies, however, can only, with some difficulty, deny that there are not myriad very well-attested phenomena which, as Léon Chertok liked to show, most definitely *do* exist and continue to defy efforts at explanation (blisters and/or hypnotic analgesia, for example). One might choose to call these something other than hypnotic phenomena today but to do so risks forgetting the complexities of the history of which they are a part.

Stengers is particularly wary of thinking that names don't matter all that much, that there aren't important pragmatic differences conveyed by them. Shorthand references to a (deconstructed) "X" might neatly encapsulate the indeterminacies of what it is one might be engaging with but it does make the differences that names have conveyed historically somewhat difficult to address.[26] More pointedly, perhaps, for her, it is precisely the *perplexity* that hypnosis gives rise to, the *questions* that it raises that make it matter and that can, given the appropriate attention, *oblige* us to think, and one can't do that by pretending the past doesn't exist.[27] It's what constrains her to, as her friend Donna Haraway might say "stay with the trouble" in her work more generally.[28] From this point of view, what is interesting about hypnosis, what makes the questions that it raises good to think with, is not so much the proliferating array of phenomena themselves and the gestures associated with them (which can encourage a kind of fascination), nor the way that they so frequently seem to be magicked into being by the very procedures that aim to prove or disprove their independent existence (although this is an issue of considerable importance, discussed at length here), but rather what it says about the "moderns." More precisely, there is an issue here about how hypnosis might trouble models of thinking that continue to insist that a phenomenon can only really be said to exist if it can meet the tests imposed on it in the name of an assumed universal model of scientific reference. Indeed, as Stengers shows throughout this book, it is precisely in the way that hypnosis forces us to revise our reliance on the comforting narrative arc of scientific progress, with its consoling conclusion "so that's that," "we now know," or "it's all a social construct," and so on, that some of its most valuable lessons are to be found.

# A Daughter of the Enlightenment

Stengers is a philosopher, not a historian, and it is as such that her engagement with hypnosis should be understood. Anyone reading her work expecting a full-blown historical study of the development and transformation of hypnosis will be disappointed. But that does not mean her work is not historically informed. Far from it. Nor, indeed, does its philosophical quality mean that it does not talk to practitioners in interestingly relevant ways. Historical research has, in fact, always offered her a way of *raising questions* that are precisely directed toward a more constructive engagement with practices, whether scientific or otherwise. She has worked extensively with the historian of science Bernadette Bensaude-Vincent, with whom she wrote *A History of Chemistry*, for example, and numerous of her other texts display a similarly focused engagement with the history of modern knowledge practices. Indeed, *A Critique of Psychoanalytic Reason* follows—with meticulous attention to detail—a number of key episodes in the encounter of experimental science with mesmerism, and in the relationship between psychoanalysis and hypnosis. Similarly, the essays that make up her later work on *Cosmopolitics* and its engagement with the invention of "factishes" in the history of the sciences offers a close study of the ways in which the controversies pervading these practices were addressed and decisions to follow one line of enquiry over another made by scientists. More pointedly, perhaps, her earlier, groundbreaking essay, *The Invention of Modern Science* (first published in the same year as *A History of Chemistry*) explicitly proposes a conceptually demanding account of the challenges that a philosophical engagement with the history of scientific practices needs to address, in a manner that shifts discussion away from rather sterile "demarcationist" approaches to whether something is or isn't actually science. For Stengers, in point of fact, the history of the sciences constitutes something of a crucial *test* for historical practices more generally, a test that also allows us to understand a little bit better her occasional, humorous self-presentation as a "daughter of the Enlightenment."

Humor, which is extremely important to Stengers, is not simply about poking fun. If she enjoys the caustic remarks of a classic Enlightenment thinker like Denis Diderot and has, at times, subjected arguments that scientists often make (when they are effectively presenting "science" to others) to considerable ridicule, one should note that this stance is not one that seeks to distance or detach itself from the histories with which it is bound up. Humor for Stengers is part and parcel of the construction of a specific kind of political project, a specific way of, as she puts it in *Invention*, "discussing the sciences and of producing debate with sciences," one which, to the extent it successfully produces what she calls a "shared perplexity," "effectively turns those it brings together into equals."[29] In this respect, humor is not an incidental characteristic of Stengers style but plays a pragmatically important role in her work and the problems it engages with. Sharing or *cultivating* perplexity (as she later frames it) is crucial for her, and her conception of the functioning of humor actually tells us something about the way in which she seeks to *inherit* Enlightenment thinking[30]. Indeed, when, in the case of hypnosis, she evokes a future in which one might be able to laugh at the "naïveté of what we still call 'reason,'" suggesting that we might be able to do so not as judges but as a collaborative "we" involved in an adventure, this is in the context of a precise reading of the complexities of the history of the engagement with, and dismissal of, hypnosis in psychoanalysis.[31]

To be a daughter of the Enlightenment for Stengers more generally, however, is emphatically *not* to sanction the myriad ways in which the inventions of science have worked to disqualify or eradicate other practices—the implication of scientific developments in the baleful history of Western colonialism, say, the marginalization and now destruction of "common sense" on the pretext of some presumed fundamental irrationality of the public, the polemic against other healing practices in the name of a haughtily disdainful ethics, a continuing blindness to gender, and so on. The history of the sciences is not the history of a general, more or less ineluctable, rational progress—what she refers to in the present book as a "uniform blanket." Nor is it a process to be characterized by an equally ineluctable logic of disenchantment, as in claims concerning the techno-scientifically empowered instrumental domination of the world. It is a history or set of histories that one can, in an always partial, always situated manner, through *anamnesis*, open back up to the generation of positive, rather than destructive, possibilities.[32] Here, with specific reference to the way in which hypnosis was marginalized in and by the invention of psychoanalysis, this entails "making the choice to attach oneself to what forced the stuttering, the permanent disarticulation of the disparate components that the Freudian machine effectively succeeded in articulating." Staying with the trouble, again. The challenge—a test, sure, but not an obstacle—that scientific practices pose for the work of historians (and of philosophers who work with them) is a redoutable one, and it's worth pausing on it briefly because an engagement with history is so important to this book. When we frame the work of scientists, a bit generally, as being in the "service of truth" it becomes quite easy to miss the specificities of their practice and its effects. Stengers argues that, pragmatically speaking, it is better to understand scientists as actors "in the service of history," as actors "whose problem is to transform history and to transform it *in such a way that their colleagues, but also those who, after them, will write history, are constrained to speak of their inventions as a 'discovery' that others could have made.*"[33] This is a shift in point of view that makes the challenge for historians much more palpable. It is very difficult to do the history of a practice that is constitutively concerned *with* and sometimes successful *at* producing something that aims to be incontestable.[34] Indeed, with the passing of time, as Stengers points out, this produces a difference between "what the history *of the* sciences makes [the historian] capable of questioning, and what this history has defined as incontestable."[35] For example, after Priestley and Lavoisier, claims for the existence of phlogiston become eminently questionable, while oxygen becomes something of a given. The historian's task is all the more difficult because the asymmetry that characterizes the histories of scientific discoveries (the asymmetry between winners and losers resulting from the resolution of debates and discussions that are part and parcel of the development of scientific practices) is also part of the history to which the historians themselves belong. How, as she puts it, could historians "not think, *like all of us,* that the Earth revolves around the Sun," that we live on a planet, which isn't flat, or that coughs and sneezes spread diseases? When they study the development of science, the challenge for historians is one of trying to "suspend the certainties they themselves share with their contemporaries,"[36] so as not tacitly to adopt the point of view of the winner in the history of a scientific discovery. And with the passing of time, because a discovery becomes an ingredient of more and more practices, an actor in endlessly ramifying networks of actors, the difference between something that *is* scientific and something that *isn't* becomes ever more difficult to put in question. Conspiracy theorists who doubt there was ever a lunar landing probably don't doubt that the Earth is round. It is difficult to unsee images of the planet taken from space.

Crucially for Stengers, being a daughter of the Enlightenment means being able to adopt an engagement with the history of the sciences that acknowledges, along with Bruno Latour (with whom she was in close dialogue for many years before his untimely death), that "we have never been modern." Indeed, this now rather well-known and quite widely misinterpreted claim exemplifies what Stengers understands by humor. For if, as she reminds us, Latour could only develop his claims about "the moderns" by focusing on the social history of the construction of scientific knowledges, this is because only a historian who knew what it meant to *have been* modern could produce such a history "*without for all that* denouncing what he had been, or unveiling the trickeries and illusions of which he had been the victim; that is, without opposing to the truths constructed by the sciences another truth with a greater power."[37] For Latour and for Stengers, the crucial interest of social histories of scientific knowledge such as Steven Shapin and Simon Shaffer's *Leviathan and the Air Pump* (which is central to Latour's book *We Have Never Been Modern*[38]) lies in the way in which they trace out a history that is at one and the same time a history of *both* the emergence and development of increasingly complex networks of nature/culture hybrids—such as Boyle's air pump—*and* the critical "purification" of the world into stark modern oppositions of the nature versus culture, fact versus value, science versus politics kind. To show, specifically, how the eminently modern distinction between the realm of the facts of nature and politics emerges, Shapin and Shaffer are obliged to show, quite precisely, how these apparently categorically distinct realms do not cease, in fact, to become blurred. Historians who succeed in *following* the workings of such processes of "mediation" (as Latour calls them) display, in principle at least, what Stengers sees as the "capacity to recognise [themselves] as a product of the history that [they are] seeking to follow the construction of." This, more broadly, is what Stengers calls humor and it entails what she calls a philosophical "art of immanence." For while humor, construed in this manner, implies that there is no (transcendent) position from outside the history of a scientific discovery one might be attempting to construct from which one could reasonably judge it, this does not preclude the possibility of problematizing this history and the situation of dependence vis-à-vis the distinctions between science and non-science, science and opinion, it has produced. The impossibility of standing outside the complex and highly ramified infrastructures, the more or less stabilized actor-networks, which are bound up with these distinctions, which have produced us and have often also had highly undesirable consequences, doesn't mean one cannot question them.[39] Indeed for Stengers as a daughter of the Enlightenment, opening up politics beyond the fatal incoherence of the nature-culture split Latour uses to frame his account, depends on doing so.[40]

In respect of this situation then, and especially in respect of her earlier work, Stengers philosophical practice entails an endeavor to bring science into politics through the careful construction of a relationship to scientific practices (in actuality or historically) that doesn't have to take refuge in "convenient arguments" that would have the effect of dismissing the uncertainties that pervade them, falling back on global distinctions (like that of nature and culture) that a closer reading of history shows to be untenable, negating or reducing the hesitations that signal that there is something at work in the practice in question that is making its practitioners think. It's difficult to find humor in the public (self)presentation of much scientific practice as irreducibly opposed to "opinion," in well-meaning campaigns concerning public engagement with science, the sometimes rather self-serving appeal to "epistemological breaks" or, indeed, the heroics of Popperian falsification.

## Deterritorializing Philosophy of Science?

In the broader sweep of the modern era, it has not exactly been unusual for philosophers to take an interest in hypnosis or its antecedents. The fact of the matter is that the rather indeterminate set of phenomena that can be gathered together, albeit problematically, under the rubric "hypnosis" have been of interest to numerous philosophers over the centuries. German Idealism in particular was especially exercised by the phenomena that Mesmer and then Puységur had so dramatically brought into play. This is evident in Kant, for example, in his *Dreams of a Spirit Seer*, in the work of Fichte, or of Hegel— who deals with animal magnetism at length in the "Philosophy of Subjective Spirit" section of his *Encyclopedia* (which discusses clairvoyance, trance states, premonitions, water dowsing, among others.) The claim, in 1818, by Romantic critic Friedrich Schlegel, in a fragment originally destined for a study of magnetism, to the effect that animal magnetism "pronounces itself as an epoch-making event in inner esoteric world history"[41] is not untypical. But sometimes the interest expressed by philosophers has had a more practical element to it. Both Bergson and, even more so, William James, toward the end of the nineteenth century, had more than a passing practical familiarity. Discussing the former, Jacqueline Carroy notes that he was actually like many young professors of philosophy of his day in experimenting with hypnosis, and his engagement was sufficiently informed that an early article on simulation and hypnotism written by him attracted the interest of Pierre Janet and Joseph Delboeuf.⁴² William James's engagement was more extensively practice-based, and his *Principles of Psychology* display a detailed interest in "hypnotic" phenomena, alternating personalities, dissociation and the like, also attracting the attention of Delboeuf, with whom he entered into correspondence.⁴³ Curiously, even a pronounced skeptic like Wittgenstein seems to have exhibited at least a little interest in it. Prior to the important watershed represented by Mesmer, one can find interesting engagements with cognate phenomena in the work of other philosophers (Spinoza's discussion of sleepwalkers, Hume on miracles, etc.⁴⁴). However, since the Second World War—at least among a great many "continental" philosophers—psychoanalysis seems to have supplanted hypnosis as the practice of choice to engage with, especially when addressing "limits"—of representation, rationality, the subject, and so on⁴⁵ while among "analytic" philosophers it is typically some variant of cognitive science that predominates.

For her part, Stengers engagement with hypnosis developed at a time when a number of other Francophone philosophers were starting to take an interest in it. Borch-Jacobsen stands out in particular here—his book, *Hypnoses*, written with Eric Michaud and Jean-Luc Nancy, which treats hypnosis in terms of the "obstinate emergence" of a "question of passivity," in some ways exemplifies a quasi-deconstructive approach to the place of both hypnosis and psychoanalysis within a philosophical account of modern thought.⁴⁶ Michel Henry, whose *The Genealogy of Psychoanalysis* thematizes the central importance of *affect* as radically refractory to the philosophy of representation, became interested in it too after a little persuasion by Stengers collaborator Léon Chertok, who even succeeded in (briefly) attracting the attention of Jacques Derrida. The philosophically trained, once-Lacanian, psychoanalyst François Roustang also stands out here for his rather dramatic turn toward hypnosis, anticipated in a series of crucial articles and then books, which he started publishing around 1980.⁴⁷ Chertok for one welcomed such engagements, but not without reservations.⁴⁸

Understandably, when reference points derived from psychoanalysis have proved less fruitful, theoretically or practically, than was perhaps imagined, engaging more directly with hypnosis might be of interest to philosophers.

But here too, we need to slow down and consider a bit more carefully the way in which Stengers the philosopher engages with the questions that hypnosis raises, because her approach contrasts quite markedly with what might count as business as usual for philosophers. She doesn't think that "contextualising" hypnosis—or psychoanalysis, for that matter—in a history of ideas per se, even a philosophical one, is in and of itself especially thought provoking, particularly given the way—as so often seems to be the case—this history seems oriented toward a celebration of Freud's great "discovery."[49] Nor is she interested in "theorizing" hypnosis. The narrative resources of modern progress and its lack of humor are not far away here. It isn't the case for her either, as the present book makes abundantly clear, that it should be hypnosis, rather than psychoanalysis or cognitive science, to which we should turn to find the key to unravelling the mysteries of subjectivity, as if it were, in fact, the locus of a more authentic truth occluded by structural-linguistic or neuro-informatic technoscience. Practices that engage in the focused exploration of what, somewhat problematically, gets called "the" psyche raise potentially interesting questions about what "we" know, what "we" are capable of, but Stengers is concerned by the kind of short-circuiting that seems to take place when philosophers seize hold of psychoanalysis for its capacity to subvert the claims of "the" knowing subject (a stance encouraged by Jacques Lacan's introduction of the Cartesian subject into psychoanalysis), or celebrate hypnosis for its capacities to deconstruct the residues of a philosophy of representation more generally, or even proclaim that we now know that humans are really, merely, information-processing machines. This does not mean that the sometimes disturbing, often fascinating, features of subjectivity thereby highlighted are uninteresting for Stengers, or, indeed, that positive human science enquiry is null and void. However, the manner in which philosophers generally interrogate the findings of knowledge practices makes it difficult to construct the humorous position of a perplexity shared with the practitioner,[50] if, indeed, they acknowledge it at all (irony, in fact, tending to function as a protection against this). Engaging in a textually sophisticated deconstruction of representation and/or how psychoanalysis struggles to differentiate itself from hypnosis form perfectly respectable philosophical strategies for the critical appraisal of claims to knowledge. However, as Chertok saw it in the late 1980s, these kinds of approach are not really sufficient for addressing the specific issues that are at play in the hypnotic "rapport" between hypnotizer and hypnotized—as if dismantling a theoretical concept of subjectivity would resolve the problem of understanding the power of hypnotic suggestion to produce blisters.[51] One must, once again, stay with the trouble.

For Stengers, what this has meant is that one must learn to engage and think with the questions that *practices*—in the plural—raise. In the case of hypnosis, this has involved learning to pay a different kind of attention to the history or histories of which it has been a part, including laboratory-based studies, not giving in to seductive claims about Copernican revolutions, of finding a way to make its resurgences and transformations make us think, and think *differently*, cultivating perplexity. More generally, for her, discussing "ideas" without reference to the kinds of experience or experiment that could put such ideas to the test is not a good way to do philosophy.[52] However, the problem here is that habitual syntaxes, academic divisions of labor, and the conventions uniting practices around scientific enquiry offer too many shortcuts, too many ways *not* to confer on the phenomenon in question the attention it deserves, for this to happen.

## Thinking with Prigogine, Chertok, and Others

For Stengers, then, philosophy is not an activity of judgment, organized in terms of questions about representation and its limits, what the human subject can "legitimately" lay claim to knowing. It is a pragmatic, experimental, *situated* activity, that is to say both speculative and practical. As we have seen, it demands humor and the ability to make "unknowns" a positive, creative, element of the problems that it poses, rather than leaving them as more or less awkward points of silence in otherwise cohesive narratives, to be returned to via an apologetic version of the "we now know" refrain. And rather than operating in an implicitly or explicitly rivalrous relationship with science, Stengers has consistently sought a constructive philosophical engagement with scientific practices, one that is predicated on a conceptually nuanced appreciation of their contingent successes in generating possibilities for thinking. In *Virgin*

*Mary and the Neutrino* she reworks the account of philosophy as creation of concepts due to Deleuze but crucially, this is accompanied by an insistence on "speaking well" of other practices and, more generally, by a concern for science "in the making" that is notably absent from Deleuze's work. It is in this processual focus on the doing of practices that we can find the rationale for her insistence on thinking with others.[53]

*Thinking with* is characteristic of much of Stengers's writing. It isn't too much of a stretch to say that even her earliest book, *La Nouvelle alliance*, cowritten with Ilya Prigogine and originally published in 1979, exemplifies the importance that this relational approach to thinking between practices has for her. *La Nouvelle alliance* seeks to establish a connection between philosophy and science which reduces neither to one wherein the scientist benevolently relegates philosophical issues to a humbled second place ("we used to think . . . but now science shows us") nor one where the philosophy explains why the science is true, celebrating an "epistemological break" say ("Prigogine has indeed conclusively proved that . . .").[54] What Prigogine and Stengers aimed at could usefully be described as a kind of *symbiotic* process, in which the convergence and complementarity of scientific *and* speculative philosophical experimentation allowed for a rigorously empiricist approach to addressing precisely the *open* character of physics.[55] With Stengers, Prigogine's non-linear dynamics is not to be addressed as a finished, accomplished, break with the past, simply marking a before and after that could be narrated in terms of the onward march of scientific progress but as raising new *questions* that imply in turn a renewed philosophical attention to the history of physics and its relations with other practices. With Prigogine, who was precisely happy to explore questions beyond what professional physicists were supposed to restrict themselves to, Stengers can, in turn, raise the crucial speculative question "and what becomings might a physicist be capable of?" that is so central to her later *Cosmopolitics*.

Like her early collaboration with Prigogine and his team, so too Stengers's work on hypnosis develops as a symbiotic, experimental process of thinking with. In this case the engagement is primarily with Léon Chertok, a psychoanalyst who practiced hypnotism. Although front and center in the present book, neither Chertok nor his work is especially widely known in

Anglophone countries. Numerous of his books—in addition to *A Critique of Psychoanalytic Reason*—have been translated into English, but these publications (on, for example, painless childbirth) do not play to the gallery with dramatic statements about the psyche or the subversive value of what an engagement with hypnosis might offer.[56] Rather than participating in or lending themselves to the academic production of generalizing "theory," they are writings that, in displaying an ongoing engagement with a complex set of phenomena, exhibit quite clearly a practice of engaging with experiments or experiences that "put ideas to the test."

Although preceded by *A Critique of Psychoanalytic Reason*, it was actually Chertok's untranslated autobiography that was the ostensible focal point for his collaboration with Stengers (and the sociologist Didier Gille). Not for nothing was this book originally called *Mémoires d'un hérétique*.[57] Having been a highly active member of the resistance in Paris during the war, following irreversible exile from the region of Eastern Europe sometimes known as "Yiddishland," settled by Litvaks,[58] a region that was constantly fought over before being effectively destroyed during the war by the Nazi genocide, Chertok received professional training as a psychoanalyst (he was analyzed by Lacan, who liked to be sneery about him). He first started practicing hypnotism more or less by accident, through his encounter with a patient, Madeleine, who proved to be a talented subject (in the best tradition of magnetism) and he had quasi-surreptitious recourse to it in his practice as an analyst throughout the 1950s. In his explorations of hypnosis and its potentialities, Chertok came increasingly to insist on the radical falsity of the claims on the basis of which psychoanalysis purported to have dispensed with it. This brought him into direct and bruising conflict with the psychoanalytic establishment in France. Indeed, his experiences of the violent "fin de non-recevoir" meted out to his work, the flat refusal to engage with it, are a powerful reminder of how a professional body defends its territory, especially when founding tenets— Freud's noisily promoted and uncritically accepted "Copernican Revolution"[59] for example—turn out not to be all they are cracked up to be.

In 1971, Chertok established a laboratory for the experimental study of hypnosis. Some of the phenomena that he explored in his laboratory—hypnotic analgesia (the suggestively induced resistance to pain) for example—have been known about in one way or another for a very long time. In the United States, in the nineteenth century, magnetism was used by dentists (among others) for anesthetic purposes, records exist of surgical procedures carried out under magnetic influence in France going back to the 1820s, ditto for other procedures, for example, the removal of scrotal tumors in India (James Esdaile's use of Jar-Phoonk in colonial Bengal), and so on.[60] From the point of view of a psychoanalysis confident in its judgments about what hypnosis is, this undertaking could only be anathema, an exemplification of the kind of grotesque manipulations generated by the pretentions to positive knowledge of techno-science (or worse)—against which analysis, of course, claimed to provide the last redoubt. And from the point of view of "Science," the kind of often extraordinary phenomena that Chertok investigated are deemed of little interest and tend to be accorded a "merely" anecdotal status, because of the insistence, particularly in the field of psychology, on the reproducibility of experimental results. This insistence precludes the selective use—and cultivation—of "talented" subjects, those with a gift for engaging in a hypnotic rapport, demanding instead the randomized statistical anonymity of perfectly substitutable individuals. Discussing this aspect of his work, Stengers's notes that what has to be created with these talented subjects is something that experimenters, doctors who claim to be scientific, and psychoanalysts are "horrified by" to wit: "a singular affective bond, a form of consent that compromises the operator [i.e Chertok] and which makes them the 'co-authors' of what is produced."[61] In Chertok's laboratory, the production of knowledge is, simultaneously, a *production of existence.*

Stengers first met Chertok in 1984 and then again in 1985, with Didier Gille. When describing their encounter, they emphasize that such an encounter could only have happened because of the way in which each flouted the expectations that the other had of them. Gille and Stengers had worried that Chertok would be a "visionary" expecting an epistemologist to confer on his discoveries the honorific title "science." For his part, Chertok feared that the interest Gille and Stengers had in hypnosis was one that would indeed be conditional on its being accompanied by a suitably explanatory theory. Chertok, however, did not have a grand theory of hypnosis, nor did he seek one, any more than it was something Gille and Stengers were interested in.[62] Moreover, Chertok did not even claim to have accomplished any special innovations with his laboratory, for example. Rather, it seems, he sought merely to be a bit more lucid about what is not known, about the "non-knowledge" of those practices— psychoanalysis in particular—that are so confident in their judgments, not just about hypnosis, but about "the" unconscious (Freud's unconscious) more generally. What mattered to Chertok—and here we can see just how important he has been to Stengers's work more generally—was the perplexity that his engagement with hypnosis produced and which, being at play throughout his work, *made him think*.[63]

Biographies are typically for winners in the history of science and, given received wisdom concerning Freud's Copernican revolution, hypnotists don't generally fall under that category. But the history of hypnosis is not a "typical" history and, as with other practices that can be grouped around the concerns of the human sciences, the singular quality of the relations that are involved in it, the "productions of existence" it entails, really matters.[64] Would the facts of Ilya Prigogine's life story affect the way in which the far from equilibrium systems he experimented with responded to their environment? No. Would a patient, by contrast, be affected by the singularities of an analyst's make up and the more or less certain body of knowledge that informs their posture (literally) in a consulting room, by their culture? For sure.[65] In the *Mémoires* Chertok's life, its complex trajectory, the risks that this courageous member of the resistance became more or less comfortable exposing himself to, is brought into intimate proximity with the development of an efficacious therapeutic practice decidedly lacking in the kinds of convenient arguments that are, for Stengers, the hallmark of the modernist dismissal of unknowns. It's difficult to avoid the risks of exposure—to contestation, to failure or worse—without the putative guarantees provided by honorific labels like "science" or "theory." So how Chertok succeeded in conferring on hypnosis the capacity to make him think, in the face of institutions, theories and practices that more or less explicitly did exactly the contrary, and negotiated the tensions between lucidity, perplexity, and other affective "assignations" is itself a lesson from which there is much to learn. The *Mémoires* in turn charts a life without purporting to offer explanations; in fact, the form of the autobiography is itself affected here with a kind of humorous treatment that can also yield a sharing of perplexity. The troubling of roles and the flouting of the expectations associated with them, the paradoxical coexistence of different kinds of affective dispositions, and an eschewal of the habit of passing judgment all seem to be thematized in and conveyed more generally by the construction of the *Mémoires*. The book is an autobiography, but cowritten, with a dialogue that is not quite a dialogue. There are multiple other voices, including numerous witnesses who flesh out Chertok's recollections, and it's not always clear who exactly is speaking. "The doctor" as Chertok is referred to throughout, becomes a sort of "intercessor," to borrow Deleuze's expression, provoking numerous questions on the part of his collaborators, and his perplexity vis-à-vis hypnosis itself seems to form a crucial narrative operator, stretching across the book and, in particular, the forty years that separate his first use of hypnosis, with Madeleine in 1949, to a transdisciplinary conference devoted to hypnosis in Paris in 1986.

## Nothing to See Here: Mesmer

As well as addressing hypnosis, both the *Mémoires* and Stengers's other crucial book with Chertok, *A Critique of Psychoanalytic Reason*, inform her early engagements with the history and philosophy of science, particularly in *The Invention of Modern Science*. They do so in different ways but exemplify her humorous "art of immanence." And both provide her with important points of reference for a constructive and demanding conceptualization of the powers of fiction beyond the "it is merely fiction" sustained by the ill-considered extension of the demands of scientific "proof" to the engagement with responsive, imaginative, not indifferent, human bodies.[66]

It is, in fact, precisely by means of a reference to the more detailed engagement with an episode in the history of hypnosis to be found in *A Critique* that Stengers can insist more generally on the problematic functioning of a model of scientific rationality, derived from the natural sciences (chemistry, to be precise) vis-à-vis the way that bodies are called upon to testify for the pertinence of scientific categories and for the position that "the" public has been expected to adopt vis-à-vis the discoveries of science ever since. The episode in question concerns Anton Mesmer, and the establishment, in 1784, in prerevolutionary France, of two commissions, one by King Louis XVI, one by the Baron de Breteuil, acting on the king's orders, to investigate his practice of animal magnetism, usually considered a key precursor of hypnosis. While Mesmer is also considered in the present book, it is worth discussing Chertok and Stengers's reading of the 1784 episode in a little detail, not in the least because, as she has argued, the taking in hand of Mesmer forms "an inaugural scene," "the first act of a two-century history" concerning the politics of Western knowledge practices,[67] and it makes dramatically evident not just the problems that arise in extending a specific model of scientific rationality to bodies and their health but also some of the mechanisms by which publics are put in their place vis-à-vis what might matter or be of importance to them. In this respect it offers an excellent example of what engaging with the stuttering history of hypnosis might offer.

A contemporary and friend of the Mozart family, Anton Mesmer had arrived in Paris in 1778 from Vienna, where he had endeavored, with a little success but also some controversy, to generate interest in and scientific recognition of his practice of animal magnetism. Dogged by rumor and innuendo regarding his therapeutic engagement with one Mademoiselle Paradis, Mesmer moved to Paris where, unlike the rather indifferent Vienna, interest in his work really took off.[68] Indeed, his practice became so popular that to cope with the numbers of patients visiting him, he devised an apparatus, the aforementioned "baquet"—a wooden tub that contained iron filings and "magnetized" water, from which a number of iron rods protruded—so that groups of patients, linked together around it by a rope, could be treated simultaneously.[69] The use of the baquet extended Mesmer's operating through "passes"—the moving of his hands down the patient's body, pressing a patient's knees tightly within his own while staring into their eyes, the use of magnetized trees, and so on. Gathered round the baquet patients would exhibit a variety of different responses to the process Mesmer induced, involving this apparatus, a rather dramatic mise en scène (music, purple robes, etc), mesmeric "passes"—touching specific parts of the body—and so on. Fainting, hysterical laughter, convulsive "crises," sometimes approximating orgasm, as well as more modest yawning, lethargy, or indifference, were all to be observed round the baquet.

Unsurprisingly perhaps, Mesmeric practice had its detractors, not in the least because of the claims that Mesmer made to science.

Mesmer and his followers ascribed the variable effects that animal magnetism involved to the circulation of an invisible magnetic fluid, in consonance with a model of scientific reference that was widespread at the time. Sickness, they argued, resulted from blockages in the circulation of this fluid, and it was precisely the work of the "passes" and of the baquet to set it back in motion. Animal magnetism was a roaring and, for the scientific establishment and the king evidently, somewhat worrying success.[70] Robert Darnton, in his book *Mesmerism and the End of the Enlightenment*, has painted a vivid picture of this period in French history, charting the enormous growth in popularity of Mesmer's work, its connections with the science of the day and the ways in which it, sometimes for better or sometimes for worse, captured the imagination of the literate French public.

> Science had captivated Mesmer's contemporaries, by revealing to them that they were surrounded by wonderful, invisible forces: Newton's gravity, made intelligible by Voltaire; Franklin's electricity, popularized by a fad for lightning rods and by demonstrations in the fashionable lyceums and museums of Paris; and the miraculous gases of the Charlières and Montgolfières that astonished Europe by lifting man into the air for the first time in 1783. Mesmer's invisible fluid seemed no more miraculous, and who could say that it was less real than the phlogiston that Lavoisier was attempting to banish from the universe, or the caloric he was apparently substituting for it.[71]

Evidently there was a strong political dimension to these developments. Darnton slyly notes that the French public took far greater interest in Mesmerism than they did in Jean-Jacques Rousseau's then recently published *Social Contract*, but he also underscores the emergence of a radical strain in the Mesmerist movement, represented by the figures of Nicolas Bergasse and Jacques-Pierre Brissot, which found in animal magnetism a useful means for developing egalitarian political claims. Indeed, as Chertok and Stengers point out, one implication of Mesmer's theory of an invisible magnetic fluid was that *all* subjects, men, women, rich, and poor, would be affected by it, regardless of social convention. In this respect at least, round the baquet, all would effectively be equal.[72]

With the benefit of more than two centuries of hindsight, Mesmer, with his purple robes and glass harmonica, to say nothing of the magnetized trees and other elements of his practice, appears to us as almost the archetype of the charlatan or quack. But given the more general popularity of scientific ideas at the time, while we may now be quick to dismiss him, hypothesizing the existence of an invisible fluid to explain what happened to people around the baquet can hardly be seen as unreasonable. Framing it in these tolerant terms—so as to make something that we find difficult to accept now appear reasonable in its day—is tempting. Contextualizing approaches that seek to show how what Mesmer was doing made sense in terms of more widely circulating scientific ideas of the time seem, generously, to afford the charlatan the benefit of the doubt. True, Mesmer may have *believed* in the existence of an invisible fluid but after all who didn't back then? Still, we do now *know* better and are not taken in by such dubious claims: times change. This, however, is not the position that Chertok and Stengers wish to adopt. Hermeneutics, in this case at least, is humorless, lacking in perplexity. Appealing historically to what made sense in the Enlightenment-enthused European culture of the time more generally as a way to address animal magnetism would subtly deprecate the arguments made by the Mesmerists and allow the singularity of the situation that obtained vis-à-vis the commissions to be magicked away, occulted by the feeling of uniform progress that Science conjures up and which the hermeneut tacitly requires.

Consistent with the more general discussion in *The Invention of Modern Science* regarding the challenges of doing the history of science, in *A Critique* Chertok and Stengers thus invite us to pay closer attention to the arguments that were made by the protagonists involved in this episode—members of the commissions, Deslon (the not entirely equivalent substitute for Mesmer in the commissions' investigations), and the botanist Jussieu, the sole dissenting voice on the commissions, in particular,[73] as a way to avoid falling into the trap, made so readily available—in this history—by psychoanalytic appeals to "Copernican Revolutions" or by cognitive-scientific claims about information processing. What transpires from the analysis that they propose of the Mesmer episode, emphasizing the shaky nature of the commissions' conclusions, the perfectly reasonable resistance opposed to them by Deslon, and the marginalized alternative approach to studying animal magnetism proposed by Jussieu, is a situation with remarkable *contemporary* interest. Often lionized by narratives of science as incessant progress and frequently cited as the origin of double-blind medical testing and the placebo, the commissioners' dismissal of Mesmer's claims about an invisible fluid and their correlative assertion that it was imagination, imitation and touch that were largely responsible for the effects of the hypothesized fluid, inaugurates instead the emergence of an apparatus that effectively produces an understanding of bodies that have the capacity—irritating for science, perhaps—of "get[ting] better for the wrong reasons"

The reading that Chertok and Stengers propose focuses in the first instance on the "confrontation" between men of science—the commissioners—and the Mesmeric crisis, because it is in this confrontation that we are to find the challenges that are raised by the attempted submission of the "irrationality" of ticklish, susceptible bodies, to the diktats of a certain model of (scientific) reason.[74] The challenge that the commissions had to address was one of how to make the crises that occurred round the baquet susceptible of scientific investigation. From this point of view, the sheer multiplicity of responses of patients to the process triggered by Mesmer was effectively such for the commissioners as to be perceived as an obstacle to their investigation, making it difficult to evaluate whether or not the effects observed could indeed be attributed to the magnetic fluid. Hence, for the majority of commissioners it was uniquely the question of whether or not the universal fluid actually existed that became the focus of their investigations. To answer this question, the commissioners effectively substituted a "laboratory" situation for the uncontrollably public situation of a treatment underway, that is to say "an experimental scene in which it is the scientist who determines which questions to ask, the experiments to try, the proofs and counter-proofs to which the objects of his investigation must be submitted."[75]

Chertok and Stengers argue that in order to do this the commissioners effectively adopted an approach analogous to that developed by one of them, Antoine-Laurent Lavoisier, for the controlled study of chemical reactions.

For Lavoisier, chemistry involved the careful weighing of all the elements involved in the chemical reactions the scientist sought to establish so as to facilitate the interpretation of unambiguous facts, an interpretation purified of all those parasitical elements—prejudices, say, beliefs—that might otherwise surreptitiously shape one's understanding.[76] Naturally, in the case of Mesmerism, wherein patients gathered around a baquet, there's nothing to be weighed, no equations to write, no equivalent of Lavoisier's *Elementary Treatise on Chemistry* and its pedagogy. There are patients and their responses but the complexity of the situation in question, the multiplicity of reactions provoked, is something of a problem for controlled scientific investigation. But if patients could be exposed to magnetic treatment individually, in a controlled manner, and if one could be satisfied that they have been appropriately distanced from their imaginations, one might be able to rely on what they say. Especially if they are respectable—from the distinguished classes—in good health, of good faith, and willing to experience any effects that might be generated, even if these turn out to be mildly disagreeable.

The experimental, quasi-Lavoisierian, quality of the approach adopted by the majority of commissioners to the study of Mesmerism was one that thus tended to preclude taking too much interest in the capacities of patients. Indeed, for the majority of commissioners these capacities were largely irrelevant, bar the dispassionate service that such "witnesses" might offer to science. They eschewed the approach advocated by Jussieu (who wished to continue to investigate what happened round the baquet), preferring instead an approach that barred from the outset the idea that the sick people treated by Deslon might be addressed as the scientists' epistemic equals, susceptible of taking a legitimate and potentially illuminating interest themselves in what they experienced. For Franklin, Lavoisier, Bailly and others, the crucial point in this regard was that Mesmer's baquet did not produce legitimate experimental facts. Only when "subjects" were investigated in isolation, in ignorance of their situation, could something approaching what the commissioners might consider to be "facts" be produced. And then, from this point of view, given that the "facts" showed that individuals who were blindfolded, for example, could display the same responses regardless of whether they were mesmerized or not, the commissioners concluded that the effects Mesmer ascribed to the magnetic fluid couldn't be seen as anything other than the artifactual products of *imagination, influence,* and *touch,* of the dramatic, collective staging of crises orchestrated by Mesmer. But what the commissioners' investigations thereby put center stage, Chertok and Stengers argue, were effectively "abstract bodies," precursors of today's statistical sample populations, *abstracted* bodies, that is to say, bodies "stripped of the means that ordinarily allow living bodies to give a meaning to what they experience, so as to make them into the reliable witnesses of an action they experienced in total passivity."[77]

There is, however, also a second element to this discussion. For although the effective verdict of the commissions with regard to the invisible fluid was "nothing to see here, it's all in the imagination, move on," the verdict was not unanimous. Deslon, the magnetist who had participated in the investigations of the commissions, did not accept that the commissioners claims were in fact conclusive, mounting an at times quite acerbic refutation of the arguments that they proposed, raising precisely the thorny question of the pertinence of the model of reference their "experimental" approach established. While the commissioners were happy to conclude that imagination played a key role, this was not something they effectively addressed themselves. And in addition to this, Jussieu effectively not only expressed a "minority" opinion vis-à-vis the claims made by the commissions in their reports but also advocated a rather different approach to studying Mesmeric crises in the first place, one that insisted on engaging seriously with what happened round the baquet, the complex multiplicity of reactions that the quasi-laboratory approach the commissioners otherwise adopted avoided. The question that both Deslon and Jussieu were effectively raising—one reason among many for the continuing interest of this episode—concerns the extent to which the experimental exploration of purely physical effects determined by the momentary action of magnetism on no matter who is appropriate to investigating the matter in hand? It may indeed have been reasonable to conclude that the magnetic fluid Mesmer appealed to for the scientific authorization of his practice didn't exist. However, like the drunk searching for his keys under the streetlight— not because it's where he lost them but because it's where the light is—there is effectively a question of the relevance of the test the commissioners imposed.

What bodies are capable of was not a matter they gave themselves the means to address, really.

Acknowledging Jussieu's minority opinion and showing that Deslon's objections to the commissioners were well-founded matter to Stengers, because they show us firstly that the "controversy" did not get "settled" in the way that say the controversy about the existence of neutrinos got settled (which, it was finally proved, *do* exist, they have a mass, scientists danced in celebration and the skeptics have had graciously to accept they were wrong[78]) and secondly that a different manner of scientifically addressing the kind of phenomenon manifest in Mesmeric crises (and by extension, magnetic sleep, and later practices) is possible. Was, in fact, possible at the end of the eighteenth century, but was not pursued: forgotten, certainly, but not "repressed" from historical memory, exactly, since it's perfectly visible in the reports of the commission, discussable as such.[79] At this key moment in the development of Enlightenment thinking what Mesmerism teaches us is that bodies get better for the wrong reasons because generating facts that are commensurate with the exercise of scientific authority involves, as already noted "constructing a convenient argument that has, as if by accident, the power to dissimulate or silence a question it feels uncomfortable addressing."[80]

Stengers returns to the Mesmer episode on numerous occasions, nuancing her early analysis with Chertok as a function of shifts in the questions that she is asking and the problems she is posing.[81] When she does so in *Virgin Mary and the Neutrino* for example, in 2006, four years after the present book was published, it is to make some of the stakes of the problem it raises vis-à-vis a world in which there is an increasingly urgent need to find new ways for scientists to engage constructively with non-scientific practices visible. Discussing specifically the fear that a seducible public can provoke among scientists (the King was worried—even the Queen was thought to favor Mesmer, it was an affront to mores, Mesmerism had undesirable political implications, something paralleled vis-à-vis Lafayette in America) she highlights the way in which the episode marks the passage of a key element of successful experimentation—whether or not a claim (in this instance, Mesmer's vis-à-vis the fluid) can resist the test of competent objections (is there, or is there not any other way of explaining what is taking place here?)—from the "register of the event to the general register of judgement." The commissioners approach to Mesmeric crises and the claims Mesmer made, separating the study of the fluid out from the complex realities of what happens round the baquet, establishes a role for science in which it is, Stengers argues, "defined as being *against* opinion, [in which] the requirement of proof arms a general tribunal that, above all else, is intended to establish a pedagogic relationship with a public henceforth defined as credulous."[82] In this respect, the Mesmer episode and Stengers recurrent engagements with it offer a marvelous illustration of the deleterious effects of scientific practices when they become Science and claim an all-terrain power of judgment that effectively has the capacity to destroy the "ruminations" of common sense.[83]

## The Invention of a Modernist Practice

The extraordinary and ramifying set of developments that take place following the effective discrediting of Mesmer—Puysegur's discovery of magnetic sleep with Victor Race, a peasant in his employ, and the awareness this yielded of the generation of a specific "state" in the somnambulist (whence Braid's later term "hypnosis," which sought to characterize this as something purely physiological) and lucid sleep, the explorations of clairvoyance, the importing of Mesmer into the United States and the development of spiritualism, and so on and so forth—have not generally figured directly as part of Stengers's discussions. There is almost a strategic quality to the approach that she adopts, insofar as it is to specific points at which the history of hypnosis stutters that she applies herself. *A Critique* is exemplary in this regard, fast-forwarding from the controversy around Mesmer's baquet to Freud, Charcot and the invention of psychoanalysis—a rather different scenario, admittedly, but one that was equally dramatic and, at least at the outset, also had some rather spectacular qualities.[84]

It is difficult to engage in a discussion of hypnosis without psychoanalysis being explicitly or implicitly present, such has been the impact of Freud, specifically within the field of Western psychotherapeutic practice.[85] Hence it is worth considering, albeit briefly, aspects of the rich and nuanced reading

Chertok and Stengers offer of Freud,[86] some of which is reprised in the present book. Suffice it to say that for a number of decades—provided one didn't look into matters too closely—psychoanalysis might have been considered the winner in this conflict-ridden history, its invention having involved, but then buried, hypnosis as an interesting but unreliable technique, one that could only address symptoms rather than underlying causes, a technique that compromises the neutrality of the analyst. Freud, of course, liked to present himself as accomplishing a Copernican revolution in the study of the mind, and there have been plenty of historians of psychoanalysis (and analysts alike) who have been very happy to ratify this self-representation.[87] For Stengers, given its "will to do science" and its appeal to a rather epic narration of the progress of scientific knowledge, psychoanalysis more generally occupies an intensely problematic position in the ecology of modern knowledge practices, one that she visited on a number of occasions. Indeed, to the extent that its inception depends in part on its dismissive treatment of hypnosis, psychoanalysis counts as an exemplary form of what Stengers calls a "modernist practice," that is to say, one of a number of practices which "in order to present themselves as scientific . . . need to *disqualify* the opinions, the beliefs, of others, the nonmodern practices of which some claim to serve as rational substitutes."[88] From this point of view, questions that have often preoccupied commentators, especially philosophical ones, such as the rather tired one "is psychoanalysis a science?" do not per se have much allure for Stengers, especially when the "science" part of the question is left more or less unproblematized, as it frequently is.[89]

Chertok and Stengers propose a reading of Freud and his relation to hypnosis that focuses, initially, on the young Freud and his stay in Paris to work at the Hôpital Saltpetrière under Charcot from 1885 to 1886. Charcot, they point out, made it possible to study hysteria and hypnosis rationally and respectably as quasi-experimental, objective phenomena.[90] At least provided one overlooked the strong quotient of eroticism involved in the public displays of hysteria, or the affective dimension of a hypnotic "rapport" (already a worry in 1784) and the elective "sensitivity" of some patients more than others to the hypnotizer. While Freud was eventually rather dismissive of an understanding of the phenomenon as operative at a psychological level, entailing direct suggestion, a quasi-Bernheimian view that he later characterized as operating "per via di porre" (by adding something, as with painting), he would be a little more circumspect with regard to the simultaneously investigative and therapeutic understanding of hypnosis he derived from Charcot, which effectively allowed something to be disclosed through a method of catharsis "per via di levare"— by taking something away—as with sculpture.

Via Charcot, hypnosis evidently was clearly important for early Freud's endeavors to address the complex question of causality that hysteria raised. Yet he would quickly relinquish the idea that it could have a role to play in his practice. And it is this abandoning of hypnosis that Chertok and Stengers follow in the strategy he adopts in founding psychoanalysis. Famously, Freud and Breuer contended that hysterics suffer mainly from reminiscences. But while—in its favor—hypnosis, acting cathartically, could address this situation by enabling the recovery of memories lost as a result of trauma, in certain instances, it could, Freud realized, equally have disturbing effects. Indeed, if it sometimes brought memories to light in a potentially highly uncontrollable manner, this was because—Freud came to argue—it didn't quite get to the core of what it was, exactly, that was being relived by the patient in the doctor–patient relationship. An early episode with a patient ("Cäecilie M" or Anna von Lieben) who threw her arms around Freud's neck dramatized the matter, revealing to him the "mysterious" element that, he argued, lay behind hypnosis. Indeed, the awkward situation Freud found himself in with von Lieben paralleled the better-known difficulties that Breuer had experienced earlier with Anna O.[91] Freud came to argue that if a patient could express love in such a disturbing way, this was not because of any personal charms that Freud himself might possess (as if!) but rather because of some disguised— and henceforth repressed—memory that such an expression repeated.[92] Here we have the core of what the present book refers to as Freud's "coup": the genial invention of the "transference," through which the complex dynamics of the doctor–patient relationship and the expressions of love and hate that manifested themselves therein could be rethought "phantasmatically" as playing out old conflicts from childhood as if present.

The invention of the transference was crucial. It enabled Freud to ward off potentially rather unwelcome criticisms to the effect that he might somehow be implanting memories in his patients, as well, by contrast, as the disturbing implications for bourgeois Vienna of his "seduction theory." If memories are phantasies rather than real memories, then not only are they no longer necessarily associated with real events, they cannot be seen as the product of suggestion, at least not in any Bernheimian sense, either. With the transference, the psychoanalyst is dealing with a quasi-anthropological datum, one to which all humans were subject—the return of the repressed, rather than with the impact of a much more specific set of historical and cultural circumstances, or indeed, with the analyst's own prejudices. From this point of view, the transference, Chertok and Stengers remind us, grounds the claims of psychoanalysis to be engaging in an effectively *analytic* operation, one that links together the therapeutic treatment of a patient with the discovery of a repressed truth permeating their being, elicited through the self-same process. Qua irrepressible, what lurks in the unconscious and can effectively be glimpsed in repetitions dramatized in the analytic relationship thereby becomes the reliable witness to and for a patient's suffering. The transference sets up becoming aware of and overcoming resistance to this truth as the effective motor of the "cure," now addressed to a scientifically determinable "psychic reality" that analysis holds the key to.

As the reference to the transference suggests, central to the reading that Chertok and Stengers propose of Freud is a focus on analytic *technique*. They are not so concerned with theory, with its contents, rational or otherwise, but with what they call the "definition of the rational means able to authorise a theory."[93] Transference generates the material out of which analytic theories can be constructed. This focus also enables Chertok and Stengers to give the references to chemistry in Freud's work a more systematic and historically consistent status. In contrast to accounts of Freud that emphasize what is often seen as his rather heroic break with "positivistic" thinking or which deliberate extensively on his metapsychology, drawing out these references enables them to insist more strongly on the *practical* parallels between what psychoanalysis seeks to accomplish and the experimental manipulation and control of laboratory artifacts, initiated with Lavoisier and extending into the nineteenth-century chemistry that inherited from him. What thereby starts to become more perceptible is the connection between psychoanalysis and an ideal of an "active" rationality evident in scientific practices at the end of the nineteenth century, but somewhat overshadowed, historically, by subsequent scientific developments. This active rationality, in which "reason does not refer to the understanding of mechanisms but to their control,"[94] as Chertok and Stengers put it, is typified by the laboratory processes of chemistry, in which standardized products, instruments, and protocols enable the operational production of its objects. Following this parallel through, Chertok and Stengers argue that what the transference effectively produces—transference neurosis—can itself be understood as a sort of controllable laboratory artifact, apparently purified of the taint of suggestion associated with hypnosis, that can be analyzed (another chemical term) in the process through which it is produced and engaged with, a process that effectively links together knowing and healing.[95] This same process of standardization, they suggest, bears equally on the formation of the analyst as a professional, submitting to a protocol requiring a rigorously neutral stance.[96]

There's not the space here to follow the subsequent developments in the discussion that Chertok and Stengers offer of the development of Freud's work, from his relation to Ferenczi up to his late pessimism with regard to the therapeutic efficacy of analysis, an "impossible task," nor, indeed, of the way in which they continue to trace the troubling resonances of hypnosis in the work of some of his inheritors (Kohut, Stern, Spence, Gill, Friedman, Kubie, among others). That Freud did not present an especially credible view of the complexities of hypnosis when justifying his own discoveries is an issue discussed in the present book in particular. It is a central element of Stengers characterization of psychoanalysis as a modernist practice and something that affects in particular the analyst's own room for maneuvre, a point that she also makes here. One might also quite reasonably ask—as was the case in her essay "Black Boxes" and as others, such as François Roustang and Mikkel Borch-Jacobsen, have also asked in their own way—whether, in the absence of a more nuanced, less caricatural appraisal of hypnosis, the "will not to suggest" implied in the insistence on the "neutrality" of the active rationality of analytic practice is not itself "the motor of the most unavoidable of suggestions?"

## Curses! A Fearful Tolerance

Between her earlier work and the present book, Stengers's thinking more generally exhibits a significant shift, a shift that effectively re-situates hypnosis in a subtly different landscape. Her *Cosmopolitics* essays, which were published five or so years prior to *Hypnosis Between Science and Magic*, took up again the question of the status of psychoanalysis, as modernist practice, and did so through a discussion of the transference or more precisely "countertransference." However, where *A Critique of Psychoanalytic Reason* had explicitly focused on these issues from the point of view of the relationship between "heart and reason" in the human sciences, *Cosmopolitics* addresses the issue from a different perspective, in terms of the rather more expansively determined problematic of the ecology of practices and through a somewhat different engagement with the difficulties the transference raises.[97] The nature of the problem that she explores in these essays, while still involving a specific questioning of the human sciences, broadens out beyond a more or less exclusive concern with modern knowledge practices into a consideration of how to address "others," those who are not, who do not feel themselves to be, "recipients of the message of modernity," but who have nevertheless found themselves to have all too frequently been the target of its practices. Key to this engagement is what she calls the "curse of tolerance." The tolerance that she is concerned to address is one that explicitly or implicitly imputes an "innocence" to all those others whose ways of inhabiting the world get redefined in terms of the "great divide," the modern bifurcation of the world as it gets relayed through the equally modern distinction between "facts" and "fetishes."[98] Underpinning the imputation this entails, which others "believe" whereas we "know," is the by turns disenchanted or heroic avowal that "we" are alone in the universe, that "a bird is only a bird," sunsets merely electromagnetic radiation, or, indeed, that the unconscious is an irrepressible force that is not to be negotiated with. From this point of view, in the absence of attested scientific references, shamanic dialogues, ritual trances or the more widespread therapeutic treatment of sacrificing a chicken are really nothing but superstitious beliefs, mumbo jumbo, susceptible for redefinition perhaps as exemplars of the efficacy of "the symbolic" and so on.[99] The curse itself is addressed specifically to "anyone who thinks they are free to redefine, in their own terms, the way in which the 'other' inhabits this world, even when they are willing to tolerate them, even when they regret their own lost innocence."[100] It forms what Stengers calls a *test*, one that enables her to tackle the "abuse of power" constituted by what some human sciences might consider to be their successes.

The way in which the *Cosmopolitics* essays explore the "transition to the limit" of modern knowledge practices that the curse aims to address is complex. The essays imply a considerable "perplication" of the questions that Stengers was asking in her earlier work.[101] It's not the case for Stengers that exceeding limits is something that is to be proscribed, as with critical philosophy of a Kantian persuasion. But limits require a careful and attentive consideration of how the questions one might ask—in this instance, regarding the relevance of thinking in terms of proofs—can be transformed. How to contest the cursed tolerance of modernist practices without the surreptitious appeal to some transcendent value in the name of which one would speak? Crucially for this introduction, while the account that Stengers proposes, specifically in the final essay "The Curse of Tolerance," once again entails a focus on technique, it is also one in which hypnosis finds itself reframed. It is, instead, to ethnopsychiatry and in particular to the work of Tobie Nathan at the Centre George Devereux in Paris that she turns.[102] Thinking with Nathan (and Latour) here, what she proposes is the possibility of an engagement with healing practices that shifts definitively away from the prerogatives of the "psy" disciplines, with their putatively reliable access to the reality of the psyche, and toward a complex concept of "techniques of influence."

Techniques of influence, as elaborated in "The Curse of Tolerance" are emphatically not to be confused with the rather clichéd view of technique that might emerge from popular representations of hypnosis. But nor, more generally, is technique to be understood as authorized by some or other notion of a scientific truth that "explains" why that technique works, even in terms of references to a now rather indeterminate notion of "psychic reality" that nearly a century and half of the psy disciplines may have habituated us to.[103] While the notion of influence has itself been one of those terms that has been more or less central to the history of hypnosis and, more specifically, to the "horrors" of manipulation with which it has been associated, the treatment that Stengers proposes of it seeks to wrest it from the more or less explicit references to unqualified notions of humanity that tend to accompany it in modernist knowledge practices.[104] For Stengers, in fact, to speak of techniques of influence is not specifically to raise a question about the nature of the reality thought to "explain" influence at all but actually to raise a more speculative question about the type of *researcher* who might be able to address non-modern healers without making precisely that kind of more or less benignly tolerant appeal. In this respect, her approach to techniques of influence highlights the way in which unknowns introduce a positive complication into the problem that the ecology of practices addresses. Nathan himself, in a key reference, proposes the notion of an "influencology" as the only defensible scientific discipline for the "analysis of different procedures for the modification of the other" but it's an appeal to science that, as Stengers shows, avoids the shortcuts through which the human sciences and specifically psychoanalysis have avoided the difficulties attendant on an engagement with other cultures, and while it might involve theory, this is theory in the service of *transformation* rather than demonstration. Crucially, Nathan's account of influence is predicated on a contrast between "fright" and "anxiety," which Stengers develops extensively, precisely because fright, to put it very schematically and somewhat crudely, shifts our focus of attention, and obliges us to take seriously the terms on which non-modern practices understand therapeutic techniques as "modifying the other."

Initially, "The Curse of Tolerance" develops the contrast between anxiety and fright through a detailed engagement with the work of George Devereux, Nathan's precursor in matters of ethnopsychiatry. For Devereux, the anxiety that perturbs the psychoanalyst in the "countertransference," had pride of place in his endeavor to construct an account of the "behavioural sciences" in which observation could finally be based on reliable data. Addressing the countertransference mattered for Devereux because it offered a way of addressing observation in which scientists did not take refuge in a "method" to protect them from the deleterious consequences of their actions (Harlow's experiments on rhesus monkeys, for example), and it led Devereux to the position that psychoanalysis was "the only psychology whose exclusive and characteristic objective is the study of what is human in man."[105] It's an approach that insists on the central role of the researcher, or observer, rather than the "observed" for as he puts it "a rat experiment, an anthropological field trip or a psychoanalysis contribute more to the understanding of behaviour when viewed as a source of information about the animal psychologist, the anthropologist or the psychoanalyst, than when it is considered only as a source of information about rats, primitives or patients. In a real behavioural science the former data are basic; the latter are epiphenomena."[106] More simply: analysis of the countertransference is "scientifically more productive of data about the nature of man."[107] Devereux's inspiration was, in part, that of quantum mechanics, a reference that Stengers underlines the importance of: with quantum mechanics, there is no "raw" data, everything that might count as such is relative to a question posed (by an observer), and not all questions are good.[108] More pointedly, as Stengers puts it, "how the practitioner 'knows' or refuses to 'know,' how she is perturbed by the other affects how she 'perceives' and 'interprets' the behaviour of that other. That is why the question always arises about what the observer 'can bear knowing' about herself."[109] The key point for Devereux is that through analyzing the perturbations or disturbances of the perceptions that s/he has during analysis, the anxieties that introduce distortions into the process in which the analyst's unconscious operates, in Freud's words, like a "receptive organ [turned] towards the transmitting unconscious of the patient,"[110] the analyst can arrive at data that provide the "essence" of an observational situation. In other words, through engagement with the countertransference the analyst becomes a "reliable witness," susceptible when, as is the case with Devereux's ethnopsychiatry, addressing people from other cultures of overcoming the distortions that this situation might provoke.[111] However, if it were indeed the case that psychoanalysis, revisited by Devereux, did have access precisely to the data that could enable a claim to the "reliable witnessing" of what is human in man [sic], this would, as Stengers points out, actually put the ethnopsychiatric researcher effectively in the position of being at home *anywhere and everywhere on the planet*, able to understand non-moderns "not differently but better than they do themselves," a privileged position that ultimately avoids having to take the other seriously.

A reference to fright, by contrast, avoids the sometimes implicit or sometimes explicit assumption of the universality of the psychic reality that Devereux's approach to anxiety sanctions. Nathan himself argues that Freud turned away from the "traumatic" logic of fright, in favor of anxiety, in part because of the sheer multiplicity of external causes that references to traumatizing, frightening, encounters entail. The "endogenous" logic of anxiety makes things much more manageable. But, more pointedly perhaps, because a reference to fright would, as Stengers puts it, "have led to the conclusion that some 'other' has intruded, has influenced or modified us, possibly even caused our metamorphosis," acknowledging it would in turn imply that the psychoanalyst, or the psychoanalytic setting more generally, could be frightening in their own right. In non-modern practices, and contrasted with anxiety, the traumatic logic of the "fright" provoked by invisible beings implies a focus of attention that is directed outwards. By engaging with the invisible beings—djinn, for example—that take hold of, possess, seize a patient, the non-modern healer's enquiry implies the existence of other universes, elsewheres that house such beings. Western theories of psychic functioning explain this away as "belief" or as evidence of a regressive mentality (within a single universe), in any event as something that modern categories are best placed to judge. But as Nathan shows, presuming the "inwardness" of psychic reality as a starting point for addressing such practices precludes an appreciation of what it is the healer is doing, the very precise technical operations of their practice. Addressing the intentionality of invisible beings and conducted in the presence of a "patient," the non-modern healer is, as Stengers puts it, engaged in a process of "deciphering obligations"—what does this being want?—that not only enables the healer to produce "clinical" material that is pertinent to the interaction underway[112] but also operates to shift the direction in which the action of invisible beings is oriented, away from the patient. Such forces can, in this respect, be negotiated with but their mode of existence is refractory to knowing within modern frames of reference. The divination practiced by non-modern healers is, in this sense, to be distinguished from Western forms of clinical diagnosis. It is not focused on interpreting a symptom in an etiological framework linked to, say, the formation of drives derived from a patient's "personal history" or to a theory of neural receptors and brain chemistry. As Nathan argues, where diagnosis in modern psychopathology "welds" the symptom to the patient, non-modern healing practices effectively and sometimes durably separate them. Exploring the potential multiplicity of universes onto which a culture opens, deciphering the obligations that this brings with it, and "negotiating" with the beings put into play is a process that, in this respect, facilitates a permanent displacement of the troublesome affect that captures or possesses someone. Latour articulates this process nicely in his discussion of Nathan's work by calling the invisible beings that non-modern healing practices engage with "trans-frayeurs," or "trans-fears" in English—an expression that is practically the homonym of "transferts" (transfers, transferences) but which underlines the point that what is targeted here do not derive from a subject but have a specific mode of existence of their own.[113]

For Stengers addressing accursedly tolerant human scientific research, the key point is that thinking in terms of fright obliges the (Western) practitioner to accept that when they are engaged in a therapeutic practice, they do not effectively have access to "a psychic 'core' common to all humans." Rather, she suggests, what Devereux's psychoanalysts really find when they consult the "perturbations" of their own being with a view to generating reliable data for research is actually a cultural core, and a rather frightening one at that because, as Stengers puts it, it has successfully "produced modern humans in such a way that they recognise and deplore 'fetishes' wherever they find their manifestations."[114] Nathan and his "technical" account of healing practices, by contrast, generate the possibility of a kind of researcher whose own practice "actively prevents maintaining an opposition between modern knowledge and 'traditional belief,'" that is to say, enables them to participate in exchanges between modern and non-modern practices that are not organized around the fateful asymmetry of tolerance that had compromised Devereux's work.[115]

With Nathan's ethnopsychiatry and its engagement with non-modern techniques of influence, the matter of how the other is modified (in a therapeutic situation) becomes, as Stengers puts it "a highly sophisticated question."[116] While his research entails often detailed discussions of traditional healers (the portrait that he offers of Réunion's Madame Visnelda, for example[117])—Nathan's therapeutic practice per se is not predicated on an appeal to the authentic knowledge of a patient's culture, any more than it starts from a presumed knowledge of psychic functioning. Indeed, while the consultation apparatus that Nathan experiments with at the Centre Georges Devereux involves wide-ranging expertise on traditional healing practices, there's nothing in it that isn't, in point of fact, highly *artificial*, down to the important role of translators often required to enable consultations to take place.[118] The crucial point for Stengers is that rather than offering us an interpretation of non-modern practices that finally arrives at the truth of what they are, Nathan succeeds in conferring on such practices the capacity to make "us" think, and it is precisely around cultivating the multiple obligations that are in play in this situation that his research is organized. The apparently highly contrived nature of what is involved in non-modern practices, in this respect, presents a particularly powerful contrast with the "purifying" tendencies of Western rationality, as in the laboratory product generated by Freud.[119] As she puts it,

> if [the practitioner of ethnopsychiatry] belongs to the adventure of scientific knowledges, it is not so as to give this adventure the power to extend the procedure of proof, the mode of convocation that produces data that has the power to create consensus . . . into the worlds of healing. It is, rather, as an antidote (remedy, poison, pharmakon) against the arrogant naivety that threatens this adventure.[120]

There is much in the analysis that Stengers develops in her discussions of Nathan's work (dangerously simplified here) that resonates with her earlier considerations of hypnosis. However, with the shift in the problem framing her work, so too there is a discrete shift in the value of hypnosis for what she is doing, a shift that is explored in the present book, especially in the final chapter. For as she notes in "The Curse of Tolerance," the "probable weakness" of the functioning of hypnotic techniques, what they are capable of doing, the "magic gestures" that they employ, is that they *reduce* these gestures to a psychological reading. In this respect at least, albeit in a rather different manner, such techniques are not as distant from the psychoanalysis Stengers had contrasted them with in her work with Léon Chertok. Indeed, faced with the perplexing capacity of bodies to respond to the lure of fiction, the history of hypnosis, as well as the often highly nuanced accounts that some practitioners have offered of what they are dealing with, displays this tendency to turn to a now psychologized reference point as a manner of authority. Correlatively, the fright such techniques are susceptible of generating (on the part of the practitioner, troubled by something for which s/he has no reliable explanation) gets transformed into "the anxiety of having to assume a role liable to inspire fantasies of omnipotence." This transformation in turn offers a suggestive "explanation" as to why "the history of [hypnotic] techniques leads us to witness the perpetual return to the evocation of a psychic agency (the Ericksonian unconscious, for example) sufficiently powerful to release the practitioner from her anxiety by allowing her to attribute to this 'other' the responsibility for the effects produced."[121] Faced then with the question of the status of such an agency, the temptation thus becomes practically irresistible to frame this in terms of notions such as the potentiality of the "human as such." This is a temptation that, as she makes clear in the present book, certain practitioners of hypnosis have not been able to avoid, and it is highly problematic.[122] Curses!

However, we should avoid thinking that the note of caution that Stengers sounds with regard to the invocations of psychic agencies, and specifically implicit or explicit references to "universal human potential" means that hypnosis is irreducibly compromised, that its history should come to a halt with the failure to constitute it as a stable object of scientific knowledge, as if the latter should somehow also function as the final court of historical judgment. Or indeed that avoiding seeking authorization by objectified scientific referents, as when certain philosophizing interpretations of hypnosis constitute it as lying beyond this aspect of Western history, would suffice to avoid the curse of tolerance. As the final chapter of the present book makes clear, that is *not* the lesson to be drawn from Nathan's work. Indeed, perhaps a key lesson to be drawn from non-modern healing practices is that practitioners, including those of hypnosis, must learn to *negotiate* with the power that their practices put into play. This is a point that is made very clearly here and which Stengers also makes in the preface that she wrote to *Créer le réel*, Thierry Melchior's Ericksonian account of hypnosis and therapy:

> to negotiate signifies knowing that the therapist is put to the test by what they are addressing, that a veritable "cultivating" is required of them, one that

renders them robust in relation to a double threat of instrumentalization: either that of technique, defined as purely operative and reduced to the rank of a routine recipe, or that of the therapist themselves, whose place and mode of intervention are then legitimated by the imperatives of the trouble they are dealing with.[123]

# Conclusion

There hasn't been much direct discussion of magic in this now lengthy introduction. Modern judgments regarding non-modern practices have rather routinely made references to magic part of the more general repertoire of "tolerant" and hence insulting dismissals of other cultures. But references to magic are not exhausted by modern or, rather, modernist framings of it. Indeed, as the last couple of paragraphs have perhaps suggested, in an ecology of practices, questions of magic are inseparable from questions of technique and the operation of "reclaiming" it that Stengers account of Nathan and influence points toward. In the present book, as in several others that were written shortly after it (*Virgin Mary and the Neutrino*, *Capitalist Sorcery*, *In Catastrophic Times*), a key resource for Stengers has been the practices of North American, neo-pagan witches. It's an interest that is crucial to Stengers's concern to address techniques outside of the accursed modernist framing of them that has been highlighted here. But "thinking with Starhawk" we might say, also marks another shift in the problem that Stengers seeks to address, and the necessity of a reference to a different kind of history than the stuttering one of hypnosis. It's a history that is characterized by a much more direct engagement with activist practices of "making sense in common,"[124] and it brings into play a different set of inheritances and the history of political struggles this involves.[125] Neo-pagan witches form a crucial resource here for reasons that are explained, in part, by the final chapter of this book, and which are closely connected to the use she makes of the term "technique" in the *Cosmopolitics*. It is, she argues, precisely outside of the strictly "medical" realm that some of the most interesting experimentation with techniques—specifically around the now rather bowdlerized notion of "empowerment"—

have occurred. Such experimentation is not to be understood in psy terms—any more than an engagement with witchcraft is to be understood as a sort of turning the clock back to what happened before witches were burned at the stake, as in an endeavor to recover an otherwise lost sense of what witchcraft "meant." "Magic" is deliberately conjugated here with "artifice."[126]

When this book was originally published, its title was *L'hypnose entre magie et science*. Hypnosis between magic and science: a title that well-entrenched habits of thinking would no doubt parse in terms of an implied, classically Enlightened, historical movement. Reversing the order of expression, as is the case with this translation, *Hypnosis Between Science and Magic* permits a more ready designation of both magic, as what Stengers refers to here as "the unknown of the question that this little book attempts to characterise," and perhaps also the capacity of what hypnosis puts in play to trouble thinking precisely where it seemed most settled.

## Acknowledgments

Many thanks to April A. Knutson for her translation and to everyone who helped finally get this book into publication, especially Matthew Fuller, Olga Goriunova, and Lynne Pettinger.

## Notes

1 Isabelle Stengers, "Black Boxes, or Is Psychoanalysis a Science?" in *Power and Invention: Situating Science,* trans. Paul Bains (Minneapolis, MN: University of Minnesota Press, 1997), p. 107.

2 *Cosmopolitics I* and *II* were translated by Robert Bonono and published by Minnesota University Press in 2011 and 2013 respectively. *Thinking with Whitehead: A Wild Creation of Concepts*, translated by Michael Chase, was published by Harvard University Press in 2011.

3 Invoking psychoanalysis in its account of Stengersian humor, an early review of *The Invention of Modern Science*, for example, completely ignores the importance of hypnosis to her engagement with scientific practices.

4 *A Critique of Psychoanalytic Reason: Hypnosis as a Scientific Problem from Lavoisier to Lacan* was published in 1992. It was preceded by a part translation of *La Nouvelle*

*Alliance* cowritten with Ilya Prigogine and followed a few years later by *A History of Chemistry,* trans. Deborah van Dam (Cambridge, MA: Harvard University Press, 1996), cowritten with Bernadette Bensaude-Vincent, *Power and Invention: Situating Science,* trans. Paul Bains (Minneapolis, MN: University of Minnesota Press, 1997), *The Invention of Modern Science,* trans Daniel W. Smith (Minneapolis, MN: University of Minnesota Press, 2000).

5   Léon Chertok, Isabelle Stengers, and Didier Gille, *Mémoires d'un hérétique* (Paris: La Découverte, 1990).

6   Léon Chertok and Isabelle Stengers, *L'hypnose: Blessure narcissique* (Paris: Les Empêcheurs de penser en rond, 1999).

7   Isabelle Stengers (dir), *Importance de l'hypnose* (Paris: J. Vrin, 1994).

8   Tobie Nathan, Isabelle Stengers, and Lucien Hounkpatin, *La Damnation de Freud* (Paris: Les Empêcheurs de penser en rond, 1997).

9   Being "good to think with" does not, however, mean that hypnosis offers Stengers the means to interpret everything else. In some respects, it's the instability of the phenomenon that matters to her, not the definitional stability that would allow it to function as a key to interpret others.

10  "[A]t equilibrium molecules behave as essentially independent entities; they ignore one another. We would like to call them 'hypnons,' 'sleepwalkers.' Though each of them may be as complex as we like, they ignore one another. However, non-equilibrium wakes them up and introduces a coherence quite foreign to equilibrium" [*Order out of Chaos*].

11  As well as by Chertok, this periodicity has been noted by Adam Crabtree in *From Mesmer to Freud: Magnetic Sleep and the Roots of Psychological Healing* (New Haven, CT: Yale University Press, 1993). Crabtree associates the resurgence of interest in hypnosis in the 1980s with the high-profile scandals around multiple personalities in the United States.

12  In their survey of the pre-history of animal magnetism, Jean-Roch Laurence and Campbell Perry note the rather public nature of convulsions and describe "assistances" (of the "small" and "deadly" varieties) as "ritualised acts of violence" *Hypnosis Will and Memory: A Psycho-Legal History* (New York: Guildford Press, 1988), pp. 39–42.

13  For a useful recent discussion of some of these developments, see Adam Crabtree "1784: The Marquis de Puységur and the Psychological Turn in the West," *Journal of the History of the Behavioral Sciences* 55 (2019): pp. 199–215. Bertrand Méheust's work, which Stengers engages with here, charts out the extent of the amnesia surrounding magnetic somnambulism and draws out its broader implications vis-à-vis the compromised historiography of psy disciplines. On Vaudou see, for example, Ken Gelder, "Postcolonial Voodoo," *Postcolonial Studies: Culture, Politics, Economy* 3, no. 1 (2000): pp. 88–98 for a discussion of some of the literature. On ethnopsychiatry, see below.

14 Clark Hull's *Hypnosis and Suggestibility* (New York: Appleton-Century-Crofts, 1933) is a key reference point for the "experimental" tradition. See also PW Sheehan and C. Perry, *Methodologies of Hypnosis: A Critical Appraisal of Contemporary Paradigms of Hypnosis* (Hillsdale, NJ: Erlbaum, 1977) and chapter 12 of Laurence and Perry *Hypnosis Will and Memory* for a review. A more recent set of engagements with hypnosis, from range of positions in psychology can be found in Amanda J. Barnier and Michael R. Nash (eds.), *The Oxford Handbook of Hypnosis: Theory, Research, and Practice* (Oxford: Oxford University Press, 2012).

15 Léon Chertok, *L'Hypnose entre la psychanalyse et la biologie: Le non-savoir des psy*, 2nd ed. (Paris: Editions Odile Jacob, 2006), p. 225.

16 "[T]his 'we' does not refer to a concrete group to which one may or may not belong, but to all recipients of the message of modernity. It is a message that, as an 'order-word' is instantly applied as soon as we hear, understand, and accept that 'we' are not like others, those we define in terms of beliefs we are proud, but possibly also pained, to no longer share" Isabelle Stengers, *Cosmopolitics II*, trans. Robert Bonnano (Minneapolis, MN: University of Minnesota Press, 2013), p. 303

17 See in reference to this the discussion of stupidity in Isabelle Stengers *In Catastrophic Times*, trans. Andrew Goffey (Leuphana: OHP, 2015).In a different context, Emily Ogden has highlighted the way that the debunking dismissal of Mesmer in the United States functioned as a mask for the debunker's own ignorance. See her *Credulity: A Cultural History of US Mesmerism* (Chicago, IL: University of Chicago Press, 2018).

18 Chertok and Stengers, *L'Hypnose*, p. 3.

19 Léon Chertok and Isabelle Stengers, *A Critique of Psychoanalytic Reason* p. 203 .

20 Mikkel Borch-Jacobsen, "L'efficacité mimétique," in *La suggestion: Hypnose, influence, transe*, dir. Daniel Bougnoux(Paris: Les Empêcheurs de penser en rond, 1991), p. 182.

21 In addition to the essay on mimetic efficacy, see also Mikkel Borch-Jacobsen, *The Emotional Tie: Psychoanalysis, Mimesis, Affect* (Stanford, CA: Stanford University Press, 1993) and *The Making of Minds and Madness: From Hysteria to Depression* (Cambridge: Cambridge University Press, 2009).

22 In the entry on hypnosis written for the *Encyclopaedia Brittanica*, Martin Orne and A. Gordon Hammer characterize it as a "special psychological state with certain physiological attributes, resembling sleep only superficially and marked by a functioning of the individual at a level of awareness other than the ordinary conscious state." It is a state that is "characterized by a degree of increased receptiveness and responsiveness in which inner experiential perceptions are given as much significance as is generally given only to external reality." There is much in the language here, for example, the conception of a "state" that might, for Stengers, raise significant questions.

23 Stengers argues that those—like Freud, for a while—who put their trust in the "relation of force" (between hypnotizer and patient) that hypnosis seems to allow to

be established, are continually being both deceived by hypnosis (it seduces, tricks, them) and disappointed by it (because it doesn't provide the guarantee of being something stable enough to provide a scientific reference for their practice). See Isabelle Stengers "The Deceptions of Power: Psychoanalysis and Hypnosis," trans. Roxanne Lapidus *SubStance* 19, no 2/3: pp. 81–91.

24  The idea of *obligation* is central to Stengers conceptualization of the ecology of practices and is discussed at length in *Cosmopolitics*. See in particular "The Science Wars" *Cosmopolitics I* chapter 4.

25  Stengers, "The Science Wars," p. 79.

26  There is a striking proliferation of names at work in the history of hypnosis. Animal magnetism, mesmerism, somnambulism, magnetic sleep, phrenomagnetism, spiritism, spiritualism, pathetism, etherology, electrical biology, electrical psychology, and so on. Names and naming are pragmatically important for Stengers. As well as the present book sees the discussion of naming in Phillipe Pignarre and Isabelle Stengers's *Capitalist Sorcery,* trans. Andrew Goffey (London: Palgrave-Macmillan, 2011). One should, however, be wary of concluding from this that hypnosis is "simply" a linguistic phenomenon, as if there were something "simple" about how language works in hypnotic techniques. On this latter point, see the preface by Stengers to Thierry Melchior, *Créer le reel: Hypnose et thérapie* (Paris: Seuil, 1998).

27  I've used the verb "to oblige" here as it has a precise meaning in Stengers's work more generally. See her discussion of requirements and obligations in the chapter on "Constraints" in "The Science Wars," the first book in her *Cosmopolitics*.

28  Stengers explores her relationship to Haraway at some length in her recent *Making Sense in Common: A Reading of Whitehead in Times of Collapse,* trans. Thomas Lamarre (Minneapolis, MN: University of Minnesota Press, 2023). Symbiotic relationships are crucial for them both, in different ways, and it is, in part, this marginalization of symbiotic practices that is at play in the history of hypnosis.

29  Stengers, *The Invention of Modern Science*, p. 66.

30  On the issue of inheritance and its link to the posing of problems, see the discussion of the twelfth camel in the introduction to Stengers, *Thinking with Whitehead*.

31  Chertok and Stengers, *A Critique,* p. xxv.

32  Anamnesis is the focus (and title) of the second chapter of the third book of *Cosmopolitics*, "Thermodynamics. The Crisis of Physical Reality" (chapter 15 of the first, combined, volume), pp. 180–8. It entails an approach to the history of science that is neither archaeological nor genealogical, in which there is no "disguising" or masking of the decisions that were made *not* to continue to address the work of losers in the history of discoveries. The contrast is being made here with a Foucauldian approach to the history of knowledge. See the discussion below of the approach adopted by Chertok and Stengers to Mesmer.

33  Ibid., p. 40.

34 It is this incontestability that scientists themselves are then happy to exploit, rhetorically transforming proven facts into all-terrain judgments that magically forget crucial differences. Newtonian mechanics might explain the "lift" that makes flight possible, but a physicist would not know how to build an aircraft.

35 Stengers, *The Invention of Modern Science*, p. 41.

36 Ibid., p. 65.

37 Ibid., p. 66. Translation modified slightly.

38 See in particular the discussion in *We Have Never Been Modern* part 1. Note that for Stengers it is Galileo who constitutes the key historical reference point for the "science vanquishing opinion" binary.

39 Ibid., chapters 3 and 4 more generally on these issues.

40 The potentially catastrophic consequences of this split are, of course, now widely evident. Dipesh Chakrabarty offers an interesting reflection on the cultural circle within which historians have turned in his *The Climate of History in a Planetary Age*, ch.1.

41 Quoted in Laurie Johnson, "The Romantic and Modern Practice of Animal Magnetism: Friedrich Schlegel's Protocols of the Magnetic Treatment of Countess Lesniowska," *Women in German Yearbook* 23 (2007): pp. 10–33. There is a considerable literature exploring Romanticism—especially but not exclusively German Romanticism—in this connection. But many of these accounts parse their analyzes in terms of a prior privileging of the truth of psychoanalysis.

42 Joseph Delboeuf was a Belgian philosopher, psychologist, and mathematician who wrote extensively on hypnotism (a term about which he expressed serious doubts). The present book draws on François Duyckaerts's presentation *Joseph Delboeuf philosophe et hypnotiseur* (Paris: Les Empêcheurs de penser en rond, 1992) and makes a crucial connection with Bergson's *Time and Freewill*. See fn43 chapter 5 below.

43 In addition to his *Principles of Psychology*, the reader might refer to the discussion of the divided self in *Varieties of Religious Experience* and more particularly, his 1896 Lowell Lectures, as reconstructed by Eugene Taylor, *William James on Exceptional Mental States: The 1896 Lowell Lectures* (New York: Charles Scribner, 1983).

44 Hume's discussion of miracles in his *An Enquiry Concerning Human Understanding* relates to the Jansenist Convulsionaries at Saint-Medard.

45 No doubt in part as a result of the widespread interest in the work of Jacques Lacan.

46 According to the testimony of Borch-Jacobsen, a project he undertook with Michaud and Nancy to publish translations of Freud's writings on hypnosis was vetoed by the Freudian "establishment," Jean Laplanche in particular. See *Mémoires d'un hérétique*, p. 282.

47 See in particular François Roustang ... *Elle ne le lâche plus* (Paris: Minuit, 1980), which includes his pointed intervention "Suggestion au long cours." An article by Roustang on Michel Henry's work was translated as "A Philosophy for Psychoanalysis?" and included as an introduction to Michel Henry, *The Genealogy of Psychoanalysis,* trans. Douglas Brick (Standford, CA: Stanford University Press, 1993).

48 These developments are discussed in chapter 4 of *A Critique of Psychoanalytic Reason* and in a more biographical register in chapter 29 of the *Mémoires d'un hérétique*.

49 Bertrand Méheust has shown to what extent psychology and psychoanalysis have captured the terms on which the history of hypnosis has been discussed (when it is discussed!).

50 Not all philosophers, of course. For James in particular, we can speak of a much more developed practical engagement.

51 Humorously, Chertok points out that while it might be easy for experts to dismiss him and his perplexity, that becomes a bit more difficult when a philosopher gets involved. See the discussion of philosophers and how Chertok engaged their interest in *Mémoires d'un hérétique* chapter 29 "Unexpected allies."

52 See *Mémoires d'un hérétique,* p. 283 on the discussion of how philosophers engaging with hypnosis have been shaped by the divorce of theory and practice in psychoanalysis. On "ideas" see the discussion of Whitehead in the final chapter of the present book and the lengthier discussion in *Reactivating Common Sense*.

53 Here I think Stengers offers an interesting way of addressing Deleuze's characterization of the "plane of immanence" that a philosophy establishes in terms of events and others.

54 On the philosophy-science relationship more generally, see the discussion of Galileo in *Virgin Mary and the Neutrino: Reality in Trouble* (Durham, NC: Duke University Press, 2023), chapter 4 "The Sciences in their Milieus."

55 See the conclusion to Ilya Prigogine and Isabelle Stengers *La Nouvelle alliance* op. cit. The discussion referred to here was not included in the original translation of *La Nouvelle alliance*, but was later translated as "The Reenchantment of the World" appearing in the collection of essays by Stengers, *Power and Invention: Situating Science*.

56 A useful collection of articles by Chertok, *La relation médecin-patient,* prefaced by Stengers, was published by Les Empêcheurs de penser en rond in 2000.

57 "A Heretic's Memories." Written with Stengers and the sociologist Didier Gille (hence the parentheses around "auto") this book has recently been republished with a new and more historically illuminating title, *Une vie de combats. De l'antifascisme à l'hypnose* "A Life of Combat. From Antifascism to Hypnosis."

58 Not a nation state, lacking any ethnic homogeneity, "Litvakie" as it is rendered in French, Chertok, Stengers, and Gille point out, was a region identified as an entity

only by Jewish people, encompassing elements of East Poland, Bielorussia, Lithuania, and Latvia. In their history of the Litvaks, Minczeles, Plasseraud, and Pourchier attribute the emergence of "Litvakie" as a name precisely to Chertok. See Henri Minczeles, Yves Plasseraud, and Suzanne Pourchier, *Les Litvaks: L'héritage universel d'un monde juif disparu* (Paris: La Découverte, 2008).

59  See Chapter 1.

60  Chertok offers a brief review of the history and literature on the use of hypnosis as analgesic in the first chapter of *Le non-savoir des psy* and also in an article written for *L'Evolution psychiatrique*, published in 1976 "Douleur et hypnose" (reprinted in Chertok *La relation médécin-patient* op. cit.). On Esdaile, see Sudip Bhattacharya, *The English, Disease, and Medicine in Colonial Bengal, 1617-1847* (Newcastle upon Tyne: Cambridge Scholars, 2014).

61  Didier Gille and Isabelle Stengers, from the preface to Chertok, *Le non-savoir des psy*, p. 12. The experimental laboratory, from this point of view, then becomes a place not of demonstration but of a controlled "showing," that is to say, not something that is subject to a procedure necessary for proof ("demonstration") but rather something that is exhibited as existing beyond the fact/artifact distinction ("showing" translates the French term "monstration" here).

62  One of the interesting points of contrast between Chertok and Roustang is that the latter seemed rather frightened that without a theory, hypnosis would be destined to "no matter what," a sort of anything goes situation predicated on a cultural difference between America, which might tolerate this, and Europe. On this point, see the introduction to François Roustang, *Qu'est-ce que l'hypnose?* (Paris: Minuit, 1994).

63  The tensions between what they call different "regimes of affectation"—lucidity, perplexity, conviction, mobilization—are explored in the final section of *Mémoires d'un hérétique* and touched on briefly in *The Invention of Modern Science*.

64  "Productions of existence" is a notion that is developed in the later sections of *The Invention of Modern Science*, and it is indexed on an engagement with the work of Chertok and of the ethnopsychiatrist Tobie Nathan.

65  Although this is, in some ways, *the* issue for psychoanalysis. See on this the discussion in François Roustang, *Un destin si funeste* (Paris: Payot, 1976) chapter 4. It's an issue that Stengers addresses in a slightly different way in her comments on the production of zombies in her preface to Tobie Nathan, *Nous ne sommes pas seuls au monde* (Paris: Seuil, 2001).

66  This question of the manner in which one addresses bodies, or more precisely, those living beings—humans in particular—whose experience integrates "having a body" (the question of whether the body has a more than human signification) is of central importance to the construction of the later *Cosmopolitics*.

67 There are plenty of discussions of this episode available. Laurence and Perry offer a useful account in *Hypnosis, Will, and Memory*, as does Crabtree in *From Mesmer to Freud*.

68 Mesmer had written a thesis in 1766, *Dissertatio physico-medica de planetarum influx*, or *Physico-medical dissertation on the influence of the planets*. François Roustang offers an interesting discussion of the notion of influence, largely free of the ironic tone references to astrology often entail, in his book of the same name. See Roustang, *L'Influence* (Paris: Minuit, 1990).

69 The Musée d'Histoire de la Médécine et de la Pharmacie in Lyon holds what is thought to be the last surviving baquet. See http://phototheque.univ-lyon1.fr/user/edit_fichiers.asp?id=10928.

70 Chertok and Stengers's discussion begins by noting the particular, and particularly gendered, concern that Mesmerism raised vis-à-vis contemporaneous mores. A secret report written for King Louis noted the special susceptibility of women to Mesmeric crises and, of course, to the seductive charms of the magnetizer. It was said that the Queen herself was a habituée of Mesmeric practice.

71 Robert Darnton, *Mesmerism and the End of the Enlightenment in France* (Cambridge, MA: Harvard University Press, 1968), p. 10.

72 Social distinctions were operative in Mesmer's practice more broadly but the point here is that such distinctions were not considered relevant vis-à-vis the effects of the magnetic fluid.

73 Inventor of the the "Jussieu system" of plant classification, Antoine Laurent de Jussieu's dissenting report was published in 1784 under the title *Rapport de l'un des commissaires chargés par le roi, de l'examen du magnétisme animal*.

74 The original French title of *A Critique of Psychoanalytic Reason* was *Le Coeur et la raison: L'hypnose en question de Lavoisier à Lacan*. This title—the first part of which - referencing Pascal might be translated as "Heart and Mind" (Gregory Bateson on the "algorithms of heart and mind") but is more appropriately and directly rendered as "Heart and Reason"—flags up more directly the problematic, conflictual relationship between the rational and the "irrational" that Chertok and Stengers engage with and which is, of course, central to the stuttering history outlined in the present book.

75 In acting in this way, Chertok and Stengers note, the commissioners were effectively perfect Kantians, exercising the right to submit nature to their questions.

76 It's important not to forget Stengers's original training as a chemist. The role of Lavoisier in the history of chemistry is discussed at length in Bensaude-Vincent and Stengers, *A History of Chemistry*. Chemistry is an important point of reference in the *Cosmopolitics* essays as well as a crucial speculative resource in, for example, *Virgin Mary and the Neutrino* where it provides a basis for thinking politics in the ecology of practices.

77 *A Critique of Psychoanalytic Reason* p. 11 [translation slightly modified].

78  Stengers, *Virgin Mary and the Neutrino*.

79  And to frame the issue here in terms of "repression" would risk a fairly significant bit of question-begging.

80  From the point of view of the relationship between science and public opinion, see also Bernadette Bensaude-Vincent's discussion of this episode in *La science contre l'opinion: Histoire d'un divorce* (Paris: Les Empêcheurs de penser en rond, 2003).

81  There are, for example, discussions of Mesmer in *Doctors and Healers*, *Virgin Mary and the Neutrino*, *In Catastrophic Times*.

82  Stengers, *Virgin Mary and the Neutrino*, Chapter 7.

83  "The ruminations of common sense" was the subtitle of the first version of Stengers's recent book on Whitehead, *Making Sense in Common: A Reading of Whitehead in Times of Collapse*.

84  The spectacular quality of the "apparatuses" established first around Mesmer's baquet and subsequently, Charcot's hysteria have been highlighted by Raymond Bellour in his book *Le corps du cinema: Hypnoses, émotions, animalités* (Paris: POL, 2009). Georges Didi-Hubermann's study of the photographic archive of l'Hopital Saltpetriere, *The Invention of Hysteria*, trans. Alisa Hartz (Cambridge, MA: MIT, 2004) offers confirmation of the dramatic visibility of hysteria in the hands of Charcot.

85  Including historiographically—a point made in very different ways by Bertrand Meheust, in the "Préalables" section of *Le défi du magnetisme* and by Mikkel Borch-Jacobsen and Sani Shamisdani, *The Freud Files: An Enquiry into the History of Psychoanalysis* (Cambridge: Cambridge University Press, 2012).

86  The interested reader should refer in particular to the first two chapters of *A Critique*.

87  In their discussion of the Freud legend, Borch-Jacobsen and Shamdasani have pointed out that he wasn't the only figure making this kind of claim at the end of the nineteenth century. However, he was, perhaps, the one who succeeded in making everyone forget the controversies surrounding it. See *The Freud Files*.

88  Stengers, *Cosmopolitics II*, p. 285.

89  On this issue and the problematization it involves, see Stengers, "Black Boxes; or, Is Psychoanalysis a Science?" . On varying the terms on which problems are posed, see *The Invention of Modern Science* and, in a different vein, *Thinking with Whitehead* introduction.

90  In some respects Charcot was continuing a trend that Braid (in the 1840s) and then Liébault (in the 1860s) had begun. However, where Liébault and then Bernheim defended a psychological approach, focusing (with Bernheim specifically) on suggestion, Charcot approached hypnosis as a specifically somatic phenomenon, addressed to the body.

91 On the troubling history of Anna O, see Mikkel Borch-Jacobsen, *Remembering Anna O. A Century of Mystification* (London: Routledge, 1996).

92 The analysis that Chertok and Stengers propose of this episode in *A Critique of Psychoanalytic Reason* builds on a 1968 essay by Chertok, "La découverte du transfert" reprinted in *La relation médecin-patient*.

93 Turning to Freud's *meta*psychological discussions of the death drive, as Gilles Deleuze does, for example, to search for a transcendental principle operative in the unconscious would, from this point of view, would seem to beg the question concerning the validity of Freud's psychology.

94 Chertok and Stengers, *A Critique* [page? 61 in the French].

95 In "The Deceptions of Power," Stengers questions the Foucauldian reading of the links between psychoanalysis and power, a reading that makes the connection between analysis and the Christian confessional. For Stengers, by contrast, there is nothing especially hidden about the power that Freud aimed at with psychoanalysis. Rather, through the active rationality that he had aimed at with the invention of the transference, Freud had effectively achieved both *control* and *purification* in a manner analogous to the power sought by experimenters.

96 See in this regard their discussion of Freud on telepathy, Chertok and Stengers, *A Critique*.

97 It is actually the *countertransference* that Stengers explores in "The Curse of Tolerance," and to be more precise, the countertransference as thematized in the work of Georges Devereux.

98 The link between modern "facts" and fetishes has been addressed in, for example, Emily Apter and William Pietz, *Fetishism as Cultural Discourse* (Ithaca, NY: Cornell University Press, 1993). See also the brief discussion of Apter and Pietz in Bruno Latour, *On the Modern Cult of Factish Gods* (Durham, NC: Duke University Press, 2010).

99 This statement is not intended to be analytically precise. But on chickens, see Tobie Nathan's first essay in the book he cowrote with Stengers, *Doctors and Healers,* trans. Stephen Muecke (Cambridge: Polity, 2018).

100 Stengers, *The Curse of Tolerance,* p. 310. See in the present book the discussion of Octave Mannoni.

101 "Perplication"—a term Stengers borrows from Gilles Deleuze—is used in *Cosmopolitics* to refer to the way in which a limit "strips" questions and distinctions of their previous "tranquil differentiations." It is a term she associates with the singular risks that the critical questioning of knowledge has for her. "Perplexity," with which it is synonymous here, is not a generality, it's not part of the tribunal of judgment. Exceeding a limit is not to be denounced, but rather to entail a transformation of the questions one asks. See *Cosmopolitics* II "Transition to the Limit," p. 285.

102 In this context Stengers in fact refers to Tobie Nathan and to Bruno Latour, her "allies" in the cosmopolitical project. With Latour's work it is the concept of the "factish" that is central to the *Cosmopolitics* essays. The essay where Latour introduces this concept, *The Modern Cult of the Factish God*, is organized, in part, around a discussion of Nathan's work. The latter provides a brief account of the development of ethnopsychiatry in France in the preface to the second edition of his book *La folie des autres: Traité d'ethnopsychiatrie Clinique* (Paris: Dunod, 2013).

103 Lavoisier and chemistry figures large in Stengers account of the joining together of science and technique. See Bensaude-Vincent and Stengers, *History of Chemistry*, chapter 14.

104 In his shift away from psychoanalysis to hypnosis, François Roustang proposed an account of influence, arguing for it as an indispensable precondition for human freedom, rather than as the regrettable fettering and/or manipulation of such freedom. See Roustang, *Influence*.

105 "The Curse of Tolerance," in *Cosmopolitics II*, p. 319. It is worth noting, in passing, that for Stengers Devereux is a near "ally" for her—important for addressing "the moderns," specifically in the way in which laboratory studies of hypnosis (of the kind that want to know whether it is real or merely an artifact) deprive the practitioner of his/her brain, but not so valuable when it comes to addressing the curse of tolerance. On the former, see the discussion in *Virgin Mary and the Neutrino*, pp. 51–6.

106 George Devereux, *From Anxiety to Method in the Behavioral Sciences,*(Berlin, Boston: De Gruyter Mouton, 1967)p. xix.

107 Ibid., p. xvi.

108 See the discussion of Devereux and quantum mechanics in Stengers's preface to Tobie Nathan *Nous ne sommes pas seuls au monde* (Paris: Seuil, 2001).

109 Stengers, *Cosmpolitics II*, p. 318.

110 The reference here is to Sigmund Freud, Standard Edition 12 "Recommendations to Physicians Practicing Psychoanalysis," the importance of which is flagged by Roustang and taken up in the preface to Nathan referred to above.

111 The analyst's unconscious is, like that of the "native" he works with, "psychically undifferentiated," that is not subject to the relativistic effects of culture.

112 Nathan argues that reading coffee grounds is, in this respect, superior to the Rorschach test, in so far as, unlike the latter, it is not directed toward a hypothetical reality of the patient qua subject but constrains the healer to generate material that is relevant to that specific interaction with the patient. See *L'Influence qui guérit*, pp. 17–18 and the discussion in Stengers, *Cosmopolitics II*, p. 326.

113 Which, incidentally, he explores at some length in chapter 7 of his *An Enquiry into Modes of Existence,* trans. Catherine Porter (Cambridge, MA; Harvard University Press, 2018).

114 Stengers, *Cosmopolitics II*, p. 335.

115 Ibid.

116 See the discussion of Lucien Hountpakin's work on the Yoruba "master of the secret" *Cosmopolitics II*, p. 327.

117 Jeanne-Paule Honorine Visnelda, better known as Madame Visnelda, was a healer whose renown was such that her death in 1991 was front page news on the island of Réunion. Her practice is discussed at length by Nathan in the first chapter of *L'Influence qui guérit*.

118 On this issue, see in particular Sybille de Pury, *Comment on dit dans ta langue: Pratiques ethnopsychiatriques* (Paris: Les Empêcheurs de penser en rond, 2005).

119 As well as offering a depiction of a consultation at the Centre George Devereux in *The Modern Cult of the Factish God* Latour also draws on Nathan in his account of beings of metamorphosis in his compendious *Enquiry into Modes of Existence*.

120 Stengers, preface to Nathan *Nous ne sommes pas seuls au monde*, p. 44.

121 Stengers, *Cosmopolitics II*, p. 447.

122 See in particular the discussion of François Roustang in Chapter 5 of the present book. Roustang's work, from *Qu'est-ce que l'hypnose?* onwards is remarkable for its nuanced addressing of hypnotic practice in precisely these terms.

123 Isabelle Stengers, preface to Melchior, *Créer le reel: Hypnose et thérapie*, p. 14.

124 Isabelle Stengers, *Réactiver le sens commun: Lecture de Whitehead en temps de debacle*, p. 24.

125 See specifically "Inheriting from Seattle" chapter 1 of Pignarre and Stengers *Capitalist Sorcery*.

126 Stengers, *In Catastrophic Times*, chapter 15 "Artifices."

# 1

# *Wounds*

We sometimes hear it said that hypnosis "doesn't exist." Even Léon Chertok, who fought relentlessly for the need to listen and to take into account the way that hypnosis questions the explanatory ambitions of psychology, psychotherapy, and even somatic medicine, was not able to keep himself from wondering if the resurgent interest he was observing would not be followed by a new period of decline and disqualification. Once again, this interest would serve to describe and analyze the manner in which researchers and doctors are susceptible to being trapped, duped, fascinated. Once again, hypnosis would no longer exist except in a mode strikingly translated as "deception" in its double meaning, in both French and English. The French "déçu" is "disappointed": that in which one had placed one's confidence did not come up to one's expectations. But the English "deceived" accuses the one who betrayed a confidence of having first seduced, of having won, that confidence in order to better trick.

Another version of this story is possible, however, and it is this version that, along with Léon Chertok, I learned to take an interest in. The hesitation between the two meanings of "deception" puts the stress on what has deceived and does not question the expectation that has been disappointed. However, the fact of being disappointed might lead to a calling into question of the type of expectation associated with hypnosis. And in this case, there would certainly be a way to learn something from this deceptive story, but what we would have to learn first concerns our ambitions to achieve knowledge.

It is this second version, calling into question not hypnosis itself but what we require of hypnosis in order to recognize its "objective existence," that

Léon Chertok argued for when he described it as a "narcissistic wound."[1] In doing this, he was diverting the meaning that Sigmund Freud had conferred on the three narcissistic wounds associated respectively with Copernicus, Darwin, and himself. Freud told the story of a royal road toward a truth that is recognized by the wound it inflicts—man sees himself successively reduced to the status of inhabitant of a simple planet in an indifferent universe, to an animal produced for the same reasons as the hippopotamus or magnolia, by biological evolution, and he finally learns that he is not even the master of his conscious life. With this story, Freud was inaugurating a narrative strategy which became a classic, reprised notably by the big names in sociobiology or in the emergence of consciousness on the basis of neural functioning. And it is precisely this fearful machine of rhetorical power that Léon Chertok intended to thwart when he spoke about the narcissistic wound that hypnosis constituted for those who intend to give psychic functioning a scientific explanation.

Freudian rhetoric brings two distinct themes into communication: a reading of the history of so-called modern sciences, in this instance those sciences associated with Copernicus and Darwin, and a reading of the resistances provoked by psychoanalysis. Correlatively, the goal of this little book will be to take very seriously the thesis according to which the deception associated with hypnosis would be a "narcissistic wound" and to explore the different reading of these two themes which is thereby induced. The history of scientific knowledge will then be recounted as an adventure and no longer as an epic, and more precisely as an adventure in danger of changing its nature when it takes itself for an epic. And this danger is nowhere more evident than in the field of the so-called human sciences, including psychoanalysis. Speaking of the "non-knowledge of 'psy'-disciplines," Léon Chertok called for a lucid understanding that the true narcissistic wound was inflicted not on "men" in general but on those who adhered to the universality of the model associated with modern sciences.

Situating psychoanalysis in the prestigious lineage which includes Copernicus and Darwin, Freud had de facto fabricated a narrative in the epic mode: "Man" progressively discovers the misplaced character of the "high opinion" he had fashioned of himself. Modern sciences would be the reward for this discovery, granting access to "universal" responses, that is to say, responses addressed to all humans, interesting all humans, wounding all humans. Escaping from this epic style doesn't imply any "relativism," in the sense, for example, in which relativism would imply that astronomers have the right to continue to situate Earth in the center of the universe. The triumph that this legend associated with Copernicus but which, historically, is above all linked to the "new astronomy" of Kepler is a triumph of astronomy, with astronomers as protagonists. The new astronomy makes a difference for any question posed by an astronomer, not by "all humans." Escaping from the epic style thus implies solely that one not forget that there is no answer without a question. And the only ones who can be wounded by a new answer are those who are interested in the question which received this answer.

Who, then, did the "Copernican revolution" wound? Beyond the effects of propaganda, the price of this triumph was the "breaking of the circle" by Kepler, that is to say, the abandonment of that which was the master reference, not for humans in general, but for astronomers and Greek philosophers: the perfection of circular movement as a movement capable of explaining its own reason.[2] In the same way, the wound specifically inflicted by the Darwinian "drift" of the species affected those *and only those,* as biologist Stephen J. Gould constantly reminds us, who wanted to conflate "Man" with some general optimum on the basis of which it is possible to situate and judge the rest of the "animal kingdom."

If the work of Copernicus-Kepler or that of Darwin owe their greatness to the difference they succeeded in making for those who shared their questions, it is because those thinkers were at the same time their own most demanding and competent judges. It is in this sense that one may speak of adventure: it's their demanding questions that set those they unite on an adventure. And from this point of view, the rhetoric that Freud articulated around the theme of the wound breaks with all usages. More precisely, this rhetoric corresponds to what intervenes only when a controversy is considered closed, when a proposition may claim to have resisted all the counter-interpretations that competent critics have proposed. It is in this way that "historians of the Earth" (evolutionary geologists and biologists) can now characterize the contrarian refusal of certain "creationists" regarding the various modes of dating the age of the Earth: they are not competent to object, they just refuse to give up their beliefs.

If modern sciences merit the designation of adventure, it is because, as long as a controversy is open, everyone is required to recognize the "competence" of adversaries and the legitimacy of their skepticism. The primordial measure of the claim to truth, in the scientific sense of the word, is its ability to overcome all competent objections. In contrast, this ability to overcome all objections is precisely what is denied by the Freudian argument. Just as—and we will come back to this—the "resistance" of the patient confirms the relevance of an analytic interpretation and, as such, is the motor of the analytic process, so the resistance that criticisms of psychoanalysis testify to can be understood on the basis of the wound that the hypothesis of the unconscious constitutes. So Freud questions the competence of all those who have not first of all accepted a loss that only the experience of analysis can impose. The only one qualified to discuss analysis is thus one who has consented to analysis, and the consent itself is verified by an *adhesion*: the identification of criticism with a symptomatic resistance applies as much to dissident analysts as to external critics.

On the other hand, to speak, as Léon Chertok does, of hypnosis as a "wound" once again links it to the scientific adventure. For this wound is not inflicted on Man in general, but very precisely on those who have seen in hypnosis the royal road, thanks to which the power of science or of reason would be affirmed capable, everywhere, of validating an "objective" reference, differentiating the past and the future, purifying knowledge of anecdote, superstition, and belief.

However, it is not sufficient to state here that those who took hypnosis for the royal road were mistaken as to the path to follow. My project, in the pages that follow, is to show that the wound in question may be propagated across the ensemble of notions that claim to pose the problem of the "psyche" in a way that is, at last, "objective" and, in so doing, claim the power to capture the ensemble of practices and troubles, intelligible or not, which should be explained in terms of this psyche.

As Michel Foucault has indicated, the possibility of conceiving such attempts doubtless marks the "entry into the modern age." Until then, truth and subject were designated by their relation to what Foucault calls "spirituality": "in truth and in access to the truth, there is something which fulfills the subject himself, which fulfills or transfigures his very being."[3] Everything changes with the "Cartesian moment":

the modern age of the history of truth begins when knowledge [*connaissance*] itself and knowledge alone gives access to the truth. That is to say, it is when the philosopher (or the scientist, or simply someone who seeks truth) can recognize the truth and have access to it in himself and solely through his activity of knowing, without anything else being demanded of him and without him having to change or alter his being as subject.[4]

The story that was initiated by the Mesmerian link between therapeutic efficacy and animal magnetism, for which the status of "scientific reality" was claimed, and prolonged by hypnosis then psychoanalysis, may be presented as "complicating the Cartesian moment." One might even be tempted to say that the "wounding" truth associated by Freud with psychoanalysis is of the order of spirituality in Foucault's sense, accessible only to someone who has first been transformed by it. Jacques Lacan went further in this direction: affirming the strict contemporaneity between psychoanalysis and modern science, he proposed making the quest for a "truth" in the modern sense the pretense that constitutes the motor of analysis, the latter having as its goal a certain "fulfillment" of the subject, marked by the dethroning of the analyst as the "subject supposed to know." The analysand no longer seeks to "explain herself" and ceases to think that someone could explain to her what she is living. The fulfillment of the subject, thus having nothing to do with objective knowledge, is moreover difficult to define: according to Lacan, henceforth, the analysand has learned "to know a bit about her desire."

The success of the Lacanian operation of redefinition doubtless depends in part on the possibility that it entails of seizing hold of the ancient resources of spirituality, which casts a wry shadow over an epoch where spirituality no longer has any other meaning than the disaffection of the modern link between consciousness and truth. Furthermore the cost of this operation is small and the reward is huge, as it effectively closes down all the questions that magnetism, hypnosis, even Freudian psychoanalysis, elicit, and authorizes a radical irony regarding the ensemble of definitions of positive psychology. But its very efficacy poses a serious question, to which it behooves spiritual disciplines to respond. Can irony, and the feeling of disdainful superiority that so often accompanies it, be part of a path to fulfillment?

Whatever the case may be, Lacanian irony might seduce philosophers because it gives no other place to the adventure of modern sciences and to the type of truth these sciences envisage than one that dismisses them as a "moment" in the history of thought. Many philosophers are seduced by any proposition which relegates scientists, their triumphant rivals in the matter of the "struggle against opinion," to a more humble position. If the ambition of science belongs to the "modern," Cartesian, moment of thought, the philosopher who learns to critique this moment is in a position to pass judgment on the adventure of modern sciences.

What would Descartes be without Galileo? An excellent philosopher perhaps, but certainly not the thinker associated with the ultimately objective definition of scientific knowledge in the modern sense of the term. Alexandre Koyré recalled that Descartes criticized Galileo's laws of movement, laws which, according to the principles of his philosophy, made no sense. While the adventure of modern sciences can be glorified as an illustration of what knowledge can be—objective, rid at last of the illusions of spirituality—its key word is nonetheless "successful": Galileo *succeeded* in interrogating the movement of falling bodies in such a way that the possibility of intervening and manipulating converged with that of defining this movement as a function of the variables it obeyed. The experimental apparatus succeeded in transforming the movement of falling bodies into a reliable witness of the manner in which it must be represented.[5]

The question of success may seem prosaic in regard to the stakes associated with what Foucault calls the "modern age of the history of truth." But, from my point of view, it offers the great advantage of breathing life into a certain indeterminacy at the heart of a narration where everything seems already to have been "played out," where the hesitations, the stutterings, the uncertainties associated with the story of a "subject" capable of recognizing the truth "solely by acts of consciousness," appear insignificant when faced with the evidence that it's the only way possible. Whoever says indeterminacy says "embarking on an adventure" of thought, a sudden appearance of "and if . . . ?" where "we don't have a choice" had prevailed.

And what if, with regard to hypnosis, we took seriously the fact that, from the point of view of "the activity of knowing" for which knowledge of a scientific sort is a stake, success might not be met? The story in which such a success would have been produced, translating an effective hold, implementing a convergence between intervention and definition, is not ours here. In that story, the link between therapy and the production of an objective definition would have been self-evident, the efficacy of the first being the consequence of the fact that the event of success has actually happened, that it has begotten a practice capable of transforming what is interrogated into a reliable witness. Taking seriously the fact constituted by non-success, however, opens up the possibility of drawing other lessons from the story of hypnosis deceiving those who put their trust in it. This story could be one of the deceptions inflicted on those who, unlike many others, really have confronted the challenge of having to effectively reproduce the model of experimental sciences in the domain of so-called human sciences.

This narrative hypothesis does not mean that deceived researchers have been heroes, because most often, as we will see, it is their opponents who have made this challenge exist by showing that the propositions that were advanced did not respond to the demands of the model. The fact remains that in the history of the human sciences, the case of hypnosis is doubtless unique: a line of research has been denied the right to "give itself an object," the definition of which would guarantee a break with opinion, a strict separation between questions the definition authorizes and those opinions might be led to ask. In the case of hypnosis, what was demanded was that only when something could thus be defined as such an object could it behave as a reliable witness for its own definition.

This is why the wound of hypnosis may be led to propagate. The history of deceptions associated with it is in effect simply the counterpart of its singularity, of the fact that it was summoned to prolong the adventure of science and not to assert the authority of science. And the failure of this attempt to prolong it cannot, on the other hand, be separated from what its ambition was. It can be said that from its "magnetic" origins, what we have, since the second half of the nineteenth century, called "hypnosis" offers a kind of constant, or more precisely the terrain of a recurrent, ambition: to finally give a rational interpretation to phenomena that can be described with the general term of "trance" and which, as such, seem to designate the supernatural or magic.

In the age of Mesmer, at the end of the eighteenth century, it was first of all a question of manifestations of diabolic possession,[6] but those interested in hypnosis at the end of the nineteenth century were also fascinated by the "mystical" postures that their subjects were apt to adopt. Religious experience could in turn become part of what hypnosis "would explain." And, to the extent that ethnological knowledge expanded, it is the ensemble of "exotic" practices, of rituals that are "magic" or that convey the presence of "invisibles," that hypnosis may have seemed destined to "unify," to reproduce in an artificial way, purified, that is, of all the elements of "belief" judged to be parasitic.

The ambition was thus considerable: constituted as an intelligible and "secular" experimental model, hypnosis would, in a single stroke, permit the light of scientific rationality to dominate over the dark and fascinating continent of the irrational. It would permit the reduction to an intelligible and reproducible phenomenon of what had before been considered as evidence of the power of the supernatural, and for American experimenters of the twentieth century, of the power of the psychoanalytic unconscious. If the story had been one of success, it would have effectively dismissed as outdated the ensemble of so-called "prescientific" techniques and knowledges concerning the passions of what can be generically called the "soul."[7]

And this ambition gives its stakes to the lucidity that Léon Chertok called for when he spoke of hypnosis as a narcissistic wound and associated it with the thesis of shrinks' '"non-knowledge". It was not a question of making the "mystery of hypnosis" the universal key permitting all knowledges claiming to designate the "psyche," psychoanalysis included, to be dismissed on the same plane. It was a question of recognizing that our double incapacity—to define hypnosis and to define a technique of therapeutic intervention which would escape the hesitations that make any definition of hypnosis stutter—precludes recourse to the narrative motor of the triumphal stories that define what Foucault called "the modern age of the history of truth": previously one "believed . . ."; today we "know . . .."

Consequently, taking seriously the question of the narcissistic wound and constructing the stakes involved, becomes a challenge. Lucidity implies taking seriously what neurocognitivists or psychoanalysts or so many others choose to turn a deaf ear to when they answer "yes, I know but even so" or claim that "as shaky and simplistic as it may be, we need a theory." But like all questioning of the "true," lucidity may lead to cynical or disenchanted renouncement. The hypothesis that this little book will introduce is that, if the interest in hypnosis might once again founder, it will be because the questions that it is liable to incite will be settled too soon: those that in fact touch on the very definition (beyond hypnosis itself) of what is supposed to satisfy the norms of the scientific approach in the domain of so-called human sciences. Lucidity founders in disenchantment when it registers limits, which it restricts itself to recalling, and to repetitive denunciation everywhere that such limits are still being ignored. But when what is important is no longer to register the limits imposing a renouncement, but to think from the standpoint of constraints, obliging a transformation of questions, a new appetite may arise, free of all nostalgia for what had to be renounced. Lucidity can then be combined with a transformation of thought, marked by the possibility of new risks.

At the end of the journey which begins here, the question of knowing how to define hypnosis will evidently not have found an answer. It is the idea that we should be able to find an answer to this question that I hope will have become incongruous. And if that is the case, the question "what do we attribute or refuse existence to" will itself be considerably complicated and will perhaps have become the test on which the mode of incorporating the so-called human sciences into the adventure of modern knowledge depends. But we must start at the beginning: an overview of the attempts to define hypnosis and their failure.

# Notes

1 See Léon Chertok and Isabelle Stengers, *A Critique of Psychoanalytic Reason: Hypnosis as a Scientific Problem from Lavoisier to Lacan,* trans. Martha Noel Evans (Stanford, CA, Stanford University Press, 1992). The same theme is developed in Léon Chertok and Isabelle Stengers, *L'Hypnose, blessure narcissique* (Paris: Les Empêcheurs de penser en rond, 1990).

2 This wound will change with the Newtonian version of Keplerian ellipses: the circle is then no longer anything but a particular instance of an ellipse, intelligibility is reestablished, but a new "arbitrary" is introduced, gravitational force acting at a distance. One can say that Einstein's general relativity took upon itself the task of

absorbing this arbitrary into the metric of space-time, but the question then becomes the diversity of fundamental interactions, and it is this diversity that contemporary theories of unification try to reduce. Will it be said that Man waits, between hope and anguish, for physicists to decide, almost four centuries after Kepler, if the physico-mathematical cosmos can finally regain its Greek perfection?

3   Michel Foucault, *The Hermeneutics of the Subject: Lectures at the Collège de France, 1981-1982*, trans. Graham Burchell (New York: Picador, 2005), p. 16.

4   Ibid., p. 17.

5   See Isabelle Stengers, *The Invention of Modern Science*, trans. Daniel W. Smith (Minneapolis, MN: University of Minnesota Press, 1997). The difference that I am maintaining here between experimental success and all "philosophy of knowledge" in the modern sense found its most radical interpreter in Bruno Latour. See *On the Modern Cult of the Factish Gods* (Durham, NC and London: Duke University Press, 2010), and *Pandora's Hope: Essays on the Reality of Science Studies*(Cambridge, MA: Harvard University Press, 1999).

6   Mesmer maintained that the transition between what was at that time a perfectly acceptable technique, using magnets ("mineral" magnetism), toward the scandalous one that put "animal" magnetism into play was inspired by the "similar" effects that the exorcist Gassner obtained by using a wooden cross. This argument implies that for Mesmer and for many of his contemporaries, it went without saying that possessions and exorcists battling with demons were going to be "secularized." The real issue was thus that the metal had no right to intervene in a "finally secular" explanation.

7   The history of language has produced a precious differentiation between "soul" and "psyche." While the term "psyche" is now indissolubly linked to the modern ambition to attribute to "psychical causation only" that which, elsewhere, designates a world peopled with invisible beings, the term "soul" has remained free and may still be linked to what signals risks recognized in all cultures: "lose one's soul," "errant soul," "stolen or enslaved soul."

# 2
# *A History that Stutters*

When should we have the history of hypnosis begin? The classical narration begins with the "animal magnetism" of Franz Anton Mesmer, the term "hypnosis" (due to James Braid, credited with being the first to have envisaged the phenomenon "apart from magnetism") thus marking "progress." Abandoning the belief in a magnetic fluid that supposedly explained the power of magnetizers, or more precisely disqualifying magnetizers, defined as ignoramuses, dupes, mystics, or charlatans, marks the creation of a new science finally worthy of this name: hypnotism. The term "hypnotism," replacing that of magnetism, indicates purification, or a finally scientific taking in hand, but the "phenomenon" itself is presumed to stay the same, only with parasitic beliefs having been eliminated from it.

Today it is possible to revisit this staging, and I will return to it, thanks to the beautiful analysis of Bertrand Méheust.[1] But it is important to stress that this change of name was not the last. "Hypnosis" succeeded "hypnotism," which had not survived the failure of its promoters' claims, a change that indicates a reference to a possible "state," that of being hypnotized, without reference to a science that would define this state. To speak of the "history of hypnosis" is thus already to disregard two explanatory systems in order to concentrate on "what remains" once those explanations have been eliminated. But this remainder itself is not a "naked" reference. The term "hypnosis" implies an analogy with sleep, which goes back to the renowned magnetic sleep of Puységur and which certain typical procedures of induction keep alive, from the "you are feeling sleepy" of cartoons to the "wake up" that often concludes a session.

Other names have been proposed, and the term "hypnosis" has doubtless only survived because none of the others was capable of imposing itself, not having been associated with a generally recognized advance.

This hesitation as to the manner of naming makes the idea of a historical continuity stutter. No one doubts that the history of electricity offers such a continuity; that is why the fact that the term *electricus* signified in 1600 "distinctive of amber" poses no problem for any physicist. Here, the term is important, even if its Greek etymology is mostly elusive. As if the signifier could not be detached from the signified. Is it the "same" thing that is produced when we say "we are now in a magnetic rapport," "I am going to hypnotize you," or "I am going to lead you into a state of deep relaxation"?

Another correlative instability, of course, concerns traits that can be associated with hypnosis. For Franz Anton Mesmer, the action of the magnetic fluid first found expression in the "crisis." But his disciple Puységur induced a "magnetic sleep" and discovered the capacity that some of the patients he had put to sleep had of diagnosing their own cases, of describing the course that the pain they were suffering would take, and even of acting as clairvoyant "doctors" for other sufferers. Throughout the nineteenth century, magnetized mediums would manifest hypersensitive talents, such as making spirits and dead people speak. But the female patients of Charcot, at the Salpêtrière Hospital, illustrate the master's theory, and the specific states (lethargy, catalepsy, somnambulism) that they presented confirmed the ability of hypnosis to artificially reproduce well-defined psychopathological states. As for the subjects of Liébeault and Bernheim, unlike those of Charcot, they are "normal," though needing psychological help.[2] And they heal, while those of Freud, apparently, "resist." In short, the field is open to as many distinctions as one could wish for regarding what would be "pure" and what would be artifact, excess, or deformation testifying to the cultural or personal convictions of the "operator."

If behaviors and interpretations vary as much as does the name, how would the persona of the "operator"[3] not be affected? Franz Anton Mesmer saw himself as the founder of a science, bringing what had arisen from the supernatural into the fold of a true science, endowed at last with a "secular" object: the study of the properties of a "secular" fluid that produces a relation (*rapport*) between animals, who are endowed with animation, if not souls ("animal" magnetism, as opposed to mineral magnetism, which was well-known at that time). But this same Mesmer became, from the beginning, the archetype of the charlatan rightly denounced by clear-headed doctors. Charcot considered he was putting his female patients into an "experimental" state permitting the reproduction, and thus interpretation, of the troubles that victims of psychopathologies spontaneously present. But in the end, he was accused of having operated in the manner of a fairground ringmaster, training his patients in theatrical behavior that he related to hypnosis. As for Bernheim, who, in contrast to Charcot, intended to bring hypnotism back to the "purely psychological" problem of the action of the "moral" on the "physical," unrelated to any psychopathology, he would finally end up "voiding" this problem, that is to say, concluding that hypnosis is nothing more than the activation of a normal property of the brain, "suggestibility." When all is said and done, the hypnotist only does what each one of us does all the time, most often without realizing it: suggesting, introducing into the other's brain an idea that it accepts. The singularity of hypnosis is thus reduced to the generality of psychological, that is, suggestive, processes.

As François Duyckaerts[4] has emphasized, it fell to Joseph Delbœuf, a contemporary of Bernheim, Freud and Charcot, to add to Bernheim's conclusion that "there is no hypnotism, there are only diverse degrees of suggestibility," a substantive that restarts the problem: "of diverse degrees and *modes* of suggestibility." This, once again, changes everything because the diversity of degrees of suggestibility can be a matter for quantitative evaluation while the question of the evaluation of modes of suggestibility is an open problem.

A history that stutters is a history that "starts again" without ever permitting a clear-cut difference between "before" and "after" allowing one to speak of "progress," even though those who construct it are entirely devoted to progress. The history of experimental hypnosis, which began in 1933 with the publication of *Hypnosis and Suggestibility: An Experimental Approach* by Clark L. Hull, has been haunted by Delbœuf's question more than it has resolved it. In fact, it has been necessary to accept the evidence that the protocols of induction in which the first experimenters trusted are not just a way to induce hypnosis in a reproducible manner. These protocols are equally what guide the experimental subjects who adopt in general the "mode of suggestibility" which they understand is expected of them. The experimenters and their quantitative scales, which aimed to characterize the degree of suggestibility, itself defined as a reliable witness for hypnosis, thus contributed to "fabricating" what they believed they were measuring in a quantitative, objective way.

That is why all protocols have had gradually to incorporate an experimental constraint with the aim of trying to define a "mode of suggestibility" that would be specific to hypnosis. The experimenter who induces hypnosis has henceforth had to concern himself both with subjects who seemed to "obey" the protocol and with others urged to "simulate" this obedience. In other words, he agrees to let himself be "duped" so that the manner in which his protocol informs both the simulators and the "hypnotized subjects" about what is expected may be made evident. Any behavior that can be simulated is thus disqualified as lacking interest, because "non-specific," the hope being that a residual difference will be able to be defined when the dupery is revealed, when it is known who is simulating and who isn't.

Martin Orne, who introduced this "simulation paradigm" in 1959, thought he could identify such a "residual difference": it was a question of a serene indifference to contradiction, "hypnotized subjects" accepting without any problem, for example, the suggestion that someone was directly in front of them while they were perfectly capable of seeing that person elsewhere in the room. But this "trance logic," which would designate a mode of suggestibility specific to hypnosis, nevertheless remains debatable. For certain experimenters, all "hypnotic behavior" may be reduced to a simple "role playing" induced by the protocol, and since hypnosis cannot be differentiated from simulation, it "does not exist."[5] The accusation of simulation was, let us note, brought as early as 1829 against a female patient of Jules Cloquet who underwent the removal of a breast under "magnetic anesthesia" without flinching. We will return to this accusation since, beyond the particular problem of hypnosis, the more general question that is posed with the problem of "role playing" is one of the submission of the subjects of psychology, whether experimental or restricted to the use of questionnaires. What I wish to emphasize here is that the laboratory has not stabilized the question of hypnosis, quite the contrary, it is the experimental situation itself that has been destabilized, put in question by hypnosis. In fact, if the behavior of hypnotized subjects, far from being "purified" by the laboratory setting, integrates this setting into the composition of their "role," the laboratory (except for its boredom) is not very different as a setting from the music hall. The experimenter is no longer the one whose role is to ask questions: he "cues" the answers, without meaning to.

Therapists, like scientists, are anxious to know how roles are distributed, who is fabricating the scene. When they practice hypnosis, it is their own position that is threatened with instability. The assurance with which they think they are able to affirm "who interprets whom" or "who understands whom" is replaced with the threat of stuttering.

At the beginning of his career, Sigmund Freud seems to have experienced the instability of positions that characterize the hypnotist/hypnotized subject relation. He discovered that whoever believes himself to be the "setting the stage" could be "directed" (and thus, from his point of view, duped) by the other. Freud remained discrete on this subject, but nevertheless, he did relate a rather significant anecdote in his *Studies on Hysteria*. His patient Emmy von N. announced to him under hypnosis that she was going to be less docile than she had been before. It is this type of announcement that, a century earlier, had led Puységur to associate "magnetic somnambulism" with the sudden appearance of a "doctor," where there had been a patient, a doctor with a lucid or even extra-lucid diagnostic. Puységur had accepted that docility might not be part of what he could count on but depended on the "choice" of the patient. Freud, however, was not ready for such an inversion of the assigning of positions: right away he gave a "post-hypnotic suggestion" to Emmy von N., the realization of which was supposed to convince her of her irrepressible docility.

It is this "discomfort" of the position of the therapist who uses hypnosis that Merton Gill and Margaret Brenman evoke to explain the fact that many American analysts, who turned to hypnosis after the Second World War, decided to give it up. For Gill and Brenman, contrary to what is the rule in psychoanalysis, the therapist who uses hypnosis takes the initiative of a role where he "does" something. He is actually announcing, explicitly or implicitly, that "because of what I do, you will find yourself able to do things that you couldn't do otherwise, and unable to do things that you could otherwise do."[6]

And this announcement exposes him since he depends on the patient who may deny this new ability (see the humiliating "but doctor I am not sleeping" that posed so many problems for Freud). Or, as Puységur's example shows, the patient may well produce this ability in a mode that questions the therapist's position. And the therapist himself must resist the temptation to endorse the role of an "all-powerful father" or "miracle worker." In short, according to Gill and Brenman, the therapeutic use of hypnosis constitutes a risk that imperils the therapist much more than the patient.

Today, the idea that the hypnotist would exercise a genuine power over the hypnotized person is rejected in a rather consensual manner. That idea has become part of what, according to the canonic model of progress, people seemed to believe in the past. It would now be understood that hypnosis is always above all autohypnosis: it is the hypnotized person who confers a pseudo-power on the hypnotist; it is she who takes center stage, she who holds the key to the intelligibility of the phenomenon. The hypnotist could be anyone or anything, sometimes a simple recording and in laboratories, a researcher obeying a strict protocol, assuring the reproducibility of the cases. Hypnosis no longer attests to anything other than a common "psychological" property, and this "disenchantment" is itself a sort of guarantee of its belonging to the scientific field.

However, it is not sufficient to define hypnosis on the basis of autohypnosis in order to constitute it as a stable object. The first part of the canonic pronouncement that signals progress, "before, it was believed . . .," has been satisfied: "People believed in the power of the hypnotist, or of the fluid." That this pronouncement constitutes a caricature of the debates between magnetizers since Puységur matters little, it's the law of the genre. On the other hand, what makes it stutter is that the content of the "we now know" remains quite undetermined. In fact, the version that our story gives of the canonic pronouncement could, in a slightly naughty manner, be phrased "before it was believed . . ., now we no longer believe." And to be concrete and complete, one will have to add that the techniques of induction that correspond to this "we no longer believe" are henceforth ingredients of the phenomenon: contemporary hypnosis, and notably what is called "new hypnosis," produces witnesses who confirm with no problem the thesis according to which everything, in fact, depends on them. They could, if they decided to, remain insensitive to the induction. Everything hangs on their willingness, their collaboration.

Beyond the question, which is still the same, of the reliability of the testimony confirming the "we are no longer like the former practitioners of hypnosis" claimed by practitioners associated with the new hypnosis of Ericksonian ascendance, the problem is posed of what "autohypnosis" means. What does "auto" designate? Does it concern a conscious "psychological" subject who has decided to follow the path that the induction indicated and try the experience proposed? Or does it concern an "other," who would thereby demonstrate that "it" must be taken into account, beyond psychology?

Sigmund Freud proposed putting autohypnosis under the sign of the repetition of a primordial tie of identification. The hypnotist would thus occupy a place that the subject delivers to him, that of an all-powerful father. And for Freud this diagnostic is equivalent to a condemnation of hypnotic practices in the field of therapy. Indeed, therapy can henceforth be charged with exploiting and thus maintaining an imaginary relation of submission, of dependence, even alienation. The therapist thus becomes the "other" of the psychoanalyst, a theme widely exploited by psychoanalysis of a Lacanian inspiration.

But this accusation can be contested. The philosopher Mikkel Borch-Jacobsen has shown that Freud's diagnostic was not in fact as clear as it seemed to be. A careful reading of Freud's central text on this subject, *Group Psychology and the Analysis of the Ego*[7] allows Borch-Jacobsen to maintain that Freud is unable to define the emotional tie of "identification" such that this tie might explain in a specific way the hypnotic relation. Indeed, in Freud's text, it is thanks precisely to this tie that Freud accounts for the birth of the "Ego" itself. The subject can therefore neither represent this tie to herself nor "become conscious" of it—it has not happened to her, because "she" didn't exist before the tie. This tie could only be repeated and would recur in the transferential as well as in the hypnotic link. Mikkel Borch-Jacobsen thus concluded that Freud did not establish the possibility for the analyst of defining himself in a stable manner "against" the hypnotist: Freud may well accuse the hypnotist, but his own claim to be able to "dissolve" the transferential link on which analysis depends, and thus to combine "healing" and "realization," becoming conscious of, is unfounded.[8]

But autohypnosis can also be understood as a modification *in actu* of the relation to the self and the world. And in this case, the hypnotic link is no longer, as it was for Freud, of the order of repetition: the hypnotist doesn't "take the place" of anyone, he is the one who "shows the way," who "teaches" the subject how to attain this possibility of modification.

From this perspective, the relation between the hypnotist and his subject will have also to be described *in actu*, because in one way or another, it "indicates," in the operant sense of the term, the way to modification. The common point of the descriptions arising from this perspective seems stable: these descriptions privilege the erasure of the ordinary boundaries between hypnotist and subject, an erasure that rehabilitates the old notion of magnetic "rapport" as an interesting metaphor. It was once again Joseph Delbœuf who, with the help of a model of suggestive hypnosis similar to Bernheim's, was able to recognize the singularity of this "rapport": "Wouldn't this sympathy, which is the reason that while talking to him I am, as it were, talking to myself, make him, when he hears me, believe he is hearing his own words [. . .]? If this interpretation has some truth, it would follow that the patient, for his part, in some way hypnotizes the agent."[9]

But must the erasure of boundaries or the "sympathy" evoked by Delbœuf, be understood in "psychological" terms, that is to say, as characterizing a modified state of consciousness of the individual subject? Or on the contrary, is it not that what we call "individual consciousness" would be a derived effect, with "sympathy" being the primordial property? What seemed established in a stable manner once again begins to stutter, and this time the stuttering affects the distribution of what we deem "normal" and "modified."

Hence, for François Roustang, hypnosis is not a "modified state" of consciousness. The hypnotic "rapport" indicates the return to a state prior to psychological constitution, such as it is fashioned by social artifice and routine mobilization. The usual mode of consciousness of a subject who believes herself to be autonomous, stable, facing a stable world, is thus denounced as being mutilated, hypnosis situating itself on the side of an emancipatory return to an authentic experience, opening up the very question of life. It is thus that François Roustang proposes to call the hypnotist an "awakener" and places hypnosis in the lineage of the practices of mystic initiation that allow for the accomplishment of the experience of reuniting with "the source from which the human being draws her force."[10]

However, the erasure of the boundaries that give the psychological subject her identity does not necessarily mean reuniting with a truth beyond common consciousness. More than fifty years ago, the psychoanalyst Lawrence Kubie had already questioned "psychological" and even "psychoanalytical" categories, which claimed to define hypnosis. Nevertheless, for him it was not a question of returning to a "truth" beyond the certainties of routine consciousness but of a "creation," "an experimental reproduction of a process of natural development."[11] Hypnosis would be the artificial analog to what happens to human infants in the (pre-psychological) process wherein distinctions are created between themselves and others, between their bodies and their environment.

One might think that Kubie is repeating the Freudian hypothesis linking the question of hypnosis and the "birth of the ego," but it's nothing of the sort. For him, the creation of distinctions is a "natural" process and not "psychodynamic" (that is to say, referent to Freudian metapsychology). That is why he can speak of experimental reproduction, calling on "artificial" means, as in all experimentation. The hypnotist does not "take the place" of anyone, and induction, suggestions, monotone sensorial stimulations or "sensorial deprivation" can play an equivalent role as operators of induction. In all these cases, induction is indeed defined as disruption of the normal connection of the subject to her body, to others and to the world, a connection whose actively constructed character, "in real time," and not determined once and for all, becomes manifest.

In *Group Psychology and the Analysis of the Ego*, Freud had himself also noted the diversity of the procedures of induction, but he had concluded that "these procedures merely serve to divert conscious attention and to hold it riveted. The situation is the same as if the hypnotist had said to the subject: now concern yourself exclusively with my person, the rest of the world is quite uninteresting."[12] The psychophysiological means used (fixation on a luminous point, etc.) had no other function for Freud but to make the world appear uninteresting, that is to say, to avoid a conscious resistance of the subject to the imaginary relation that the hypnotist is proposing. For Kubie, on the other hand, induction has a "psychophysiological" effect; it leads to the relative disappearance of the border that separates the "ego" and its environment, a disappearance that he describes, in a phenomenological and not psychological mode, in terms of "reciprocal engulfing."

Unlike Freud, Kubie was endowed with the sort of humor that seems appropriate when one is interested in hypnosis: he completely accepted that his interpretation and that of Freud constitute the two branches of an undecidable alternative. Freud could relate the psychophysiological dimensions of induction back to the necessity of concentrating the attention of his patient. As for Kubie, he could interpret the psychodynamic ingredients of the induction practiced in a therapeutic framework as a means to overcome emotional obstacles that could thwart the psychophysiological transformation aroused by the induction process. In other words, *if induction must at the same time produce and overcome*, it may nourish contradictory interpretations as to what is produced and what must, for this to happen, be overcome.

However, these contradictory interpretations have completely different consequences. For Freud, what is at stake in induction—"pay attention only to me"—also explains hypnosis, defined first of all in terms of suggestibility. For Kubie, the engulfing that marks the disappearance of boundaries does not in the slightest allow an understanding of hypnosis itself. For the hypnotic state is in fact characterized by the *reappearance of an "autonomous" subject in connection with a highly differentiated world*. Once the subject is "under hypnosis," she becomes capable of responding or of refusing to respond, of discussing what she perceives or of obeying suggestions. In a word, she is no longer "captivated," a border is recreated and, with it, a (modified) relation to the self and the world.[13] Unlike the adherents of "modified states of consciousness," which must be defined as a difference between "normal consciousness" and "modification," hypnosis, according to Kubie, is thus a state of "normal" consciousness; it is the subject himself who is "modified." And Kubie confers no particular privilege on this "artificial modification." He does not place it under the sign of any "purity" but predicts on the contrary that the modified relation to the world will integrate traits stimulated by the induction procedure, that is to say, also by the manner in which the hypnotist and the subject conceive hypnosis.

The great interest of Kubie's position is thus to articulate the dominant rival interpretations of hypnosis as non-contradictory, but both partial: those that place it under the sign of a "gaping" of the subject captivated by the hypnotist and those that place it under the sign of reunion with a "truth" usually overshadowed by the certainties of consciousness. The crucial point, for him, is the tipping point between induction and "state," the difference between the moment of engulfing when the boundaries are erased and the moment of artificial recreation of new boundaries that may, if the case arises, incorporate the hypnotist (as if his words were now coming from the subject herself). If the hypnotist is interested in pathways of the initiatory type, the subject will artfully follow him along these paths.

On the other hand, Kubie's interpretation gives a positive meaning to the devastating pronouncement "hypnosis does not exist." All that exists, according to this interpretation, are techniques permitting the replaying of the identity of the "ego," that is to also say the identity of what the "ego" is dealing with. Kubie's argument is in fact a "technician's" argument: it does not rest on a representation of what hypnosis would be but refers to practice. The tipping point that he proposes conveys the knowledge of the contrast between what is not suitable "at the beginning" and what becomes possible "next." After the induction stage, if the hypnotist is not limited by a protocol that prescribes and forbids, he can indeed engage in discussion "with" his subject, which is the practical sign that she has recovered her autonomy.[14]

Today, Thierry Melchior, practitioner of a "communicational" type of hypnosis, with Ericksonian affiliations, takes up a reading somewhat similar to that of Kubie, but from the starting point of a pragmatic linguistics.[15] Like Kubie, Melchior is completely at ease with the ambiguities of hypnosis, because, for him, the distinctive feature of the induction procedure is to blur the stable differentiations that define the psychological subject, and notably the distribution between what is defined as "voluntary" and "involuntary" behavior. For Melchior, the term "hypnosis" itself is an "empty signifier," and it must be so in order to be able to "reframe" a very great diversity of signs, and to stabilize them as signs of how the behavior has become "other." And notably of how the culturally accepted definition of private sensations, sensations that no one other than the subject is supposed to share, no longer obtains. The hypnotist suggests that he has access to what the subject feels, but in a way that will not incite any opposition for it is always a question of what the subject "perhaps" feels. In fact, he transforms somewhat inevitable traits of experience into so many signs of the hypnosis that is establishing itself. But the manner in which these traits are reframed, by which the subject is led to experience as "involuntary," gestures that he thought subject to his will, is not of the order of trickery, no more so, in any case, than the categories that usually define the "private," or the difference between what would be voluntary and involuntary. The efficacy of hypnosis as an "invitation to differ" does not need any specific explanation, because so-called "normal" or "common" experience itself stems from a cultural and social framing operation. The question of the "truth" of hypnosis, of what it testifies to, is thus not even posed, and drags along with it to the dustbin of badly posed questions the ensemble of psychological categories presupposing the possibility of "objectively" identifying the (normal) mode of psychic functioning.

Nevertheless, the last word on the question of freewill has not been said. That the subject's conscious and voluntary "I can" might respond to the "you can" that has been addressed to her since early childhood, would not perhaps have surprised the magnetizers. Certainly, since Puységur, somnambulism has been associated with the "will" of the magnetizer, but Puységur had already complicated this notion of "will": it is not imposed on the patient, in the way Bernheim believed himself able to impose his suggestions; the will is what the magnetizer can "give" to the patient who is lacking it precisely because he is sick. The will is thus no longer that of the magnetizer, but something he is capable of transmitting, like a fluid. As for Delbœuf, who, unlike Puységur, did not limit his operations to the project of healing but explored what may be understood as the submission of the subject, he concluded:

> When one puts the arm of the subject in a state of catalepsy and orders him to try to lower it, he makes no appropriate effort at all. To the contrary, he moves antagonistic muscles and thus behaves as if he was not able to lower his arm. This is a simulated catalepsy, and he is duped by his own simulation. It is in this sense that I say that he lends himself compliantly to acting out catalepsy. This compliance is unconscious; it is he who, without

knowing it, wills what he was commanded to do. Hypnotism does not annihilate, it exalts the will.[16]

The dissociation of ideas of "compliance" and "submission" that is proposed here is important. Submission indicates a weakness, compliance a talent, which may be accompanied by all the lucidity one wishes. The notion of "reframing" proposed by communicational hypnosis is thus separated from what could stabilize it, the "*this is only an* operation of reframing" which gives the impression that one has understood. Reframing is proposed but the subject still needs to have the capacity, indeed the talent, that bringing this proposition into play assumes. Moreover, it is toward this position of the problem that many contemporary experimenters are oriented, having finally concluded that it is impossible to associate the susceptibility of subjects to hypnosis with a "pejorative" register—a weak, dependent, neurotic, suggestible character, and so on. In fact, it is now those who "lack susceptibility to hypnosis" who may find themselves being characterized by certain researchers as too "closed to the world," "unable to cast off the moorings," "unable to let go," in contrast to the "talented" subject, endowed notably with an intense capacity for absorption and for the creation of visual or auditory images. Correlatively, what the scales of suggestibility measure has changed meaning.[17] It is no longer a question of evaluating the degree of the subject's submission to suggestions, but his *ability* to realize them, or more precisely, to *actualize* them.

The history stutters. Hypnosis was born, in opposition to magnetism, in order to submit subjects to the norms of experimental inquiry. Experimentation rediscovers, somewhat dulled, the magnetizers' respect for the talent of their mediums.

# Notes

1   Bertrand Méheust, *Somnambulisme et Médiumnité*, vol. I. et II (Paris: Les Empêcheurs de penser en rond, 1999).

2   Today, in the words of analysts, the "subject" refers to the great philosophical tradition in which it is supposed to be inscribed and which it is supposed to subvert. But the first "subjects" were those who were expected to make psychology a science. See

Jacqueline Carroy, *Hypnose, Suggestion et Psychologie. L'inventions des sujets* (Paris: PUF, 1991).

3  Needless to say, this operator will also have different names. In this text, I have chosen the term "hypnotist," which, contrary to "hypnotizer," indicates that the one who operates is not by definition a practitioner trusting in the efficacy of hypnosis but may belong to a field defined by the question of scientificity.

4  See François Duyckaerts, *Joseph Delbœuf, philosophe et hypnotiseur* (Paris: Les Empêcheurs de penser en rond, 1992), p. 128.

5  See the debate, reuniting the principal protagonists of experimental hypnosis, around the "eliminativist" theses of Nicholas Spanos, in *The Behavioral and Brain Sciences*, vol. 9 (1986), pp. 449–502 and vol. 10 (1987), pp. 773–81.

6  Merton Gill and Margaret Brenman, *Hypnosis and Related States. Psychoanalytical Studies in Regression* (New York: International Universities Press, 1959), pp. 369–70. Gill and Brenman introduced in this book the somewhat paradoxical notion of "regression in the service of the ego" which well expresses the ambiguity of hypnosis: according to this definition, the "ego" which seems forgotten remains nevertheless perfectly capable of determining the modalities and limits of this forgetting.

7  Sigmund Freud, "Group Psychology and the Analysis of the Ego, " in *Standard Edition*, vol 18 (London: Hogarth Press, 1949).

8  On this subject, see the confrontational theses of L. Israël and of Mikkel Borch-Jacobsen in Léon Chertok, Mikkel Borch-Jacobsen et al., *Hypnose et Psychoanalyse* (Paris: Dunod, 1987).

9  Joseph Delbœuf, "Quelques considérations sur la psychologie de l'hypnotisme," republié in *Le Sommeil et les Rêves et autres textes*, edition du Corpus des Œuvres de philosophie en langue française (Paris: Fayard, 1993), p. 422.

10  François Roustang, *Qu'est-ce que l'hypnose?* (Paris: Minuit, 1994), p. 151.

11  Lawrence Kubie and Sidney Margolin, "The Process of Hypnotism and the Nature of the Hypnotic State," *American Journal of Psychiatry* 100 (1994): pp. 611–22.

12  Freud, "Group Psychology and the Analysis of the Ego," p. 96.

13  In *Le Choc des sciences psychiques*, volume 2 of *Somnambulisme et Médiumnité*, p. 324–6, Bertrand Méheust stresses that "mediums" were not by definition "asleep" or "unconscious": the "lucidity" of the visionary could be coupled with the possibility of describing and interpreting the experience.

14  The music-hall hypnotizers could equally bear witness on this point. In the first stage, their procedures aimed at both the induction of guinea-pig "hypnotizable" spectators and the elimination of "recalcitrant" others. The tipping point intervenes when the hypnotizer risks proposing to those he has kept to enter into roles that demand a certain improvisation. According to the testimony of some, the modification is marked when the question of autonomy has lost all importance. Perhaps one can

correlate the fact that the procedures of induction of the new hypnosis do not suggest to the subjects that they lose their autonomy with the testimony of subjects insisting on the absence of loss of autonomy.

15 Thierry Melchior, *Créer le réel. Hypnose et thérapie* (Paris: Seuil, 1998).

16 Joseph Delbœuf, "M. Liégeois et les suggestions criminelles," in *Le Sommeil et les Rêves et autres textes*, p. 368.

17 This new reading of hypnosis, initiated by the works of Joséphine Hilgard, led to the weighing up of a trait named "absorption." See A. Tellegen and G. Atkinson, "Openness to Absorbing and Self-Altering Experiences," *Journal of Abnormal Psychology* 83 (1974): pp. 268–77.

# 3
# *Lessons from History*

The psychology laboratory is a place where, in the name of science, people agree to become "subjects." Its obligatory prerequisite is thus consent: that of putting oneself in the service of science, and it ends up in an acceptance of submission. The subjects do not discuss the interest of the questions posed to them, the hypotheses that these questions imply, or the manner in which their answers will be interpreted. They accept the role conferred on them, that of responding to questions initiated by the scientist and for which he alone holds the meaning.

This acceptance of submission suffices to differentiate the psychology laboratory from the laboratories of physics, chemistry, or molecular biology. In these cases, "submission" is of the order of success: the event that is the production of a phenomenon responding in a reliable manner to the experimental question. With successful experimentation, the phenomenon becomes a witness for the pertinence of this question, conferring legitimacy on it. In the human or animal psychology laboratory, by contrast, a "rebel" subject or an uncooperative rat will have to be eliminated. Submission, in these laboratories, the fact of agreeing to respond, no longer vouches for the pertinence of the question. It is of the order of a condition.

However, when the required submission may be considered a "mere" condition, "separable" from the behavior being studied, it is not impossible to forget this difference: submission will then be considered a neutral condition simply allowing for the observation of a behavior that is in itself well-defined. One can thus understand why the psychology laboratory privileges what it believes it can define as an "irrepressible trait" as opposed to intentional behaviors. This, notably, is what optical and auditory illusions stage. In this case, the subject testifies "in spite of himself." And this privilege has been extended to cases where the subject responds, in an intentional manner to be sure, but to what he believes the situation to be, although in fact, he has been duped. Without knowing it, he answers a question that he didn't know had been asked.

If hypnosis has interested experimenters, it is to the degree that it could fall into these categories. The subject would certainly have consented to hypnosis, but once hypnotized, he would testify uncontrollably to something that does not designate him as an individual but which, through him, designates "the psyche" as such. It is the psyche, what affects it and the manner in which it can be affected, that would be testifying. That is why the manner in which the subject enters into the role proposed to him under hypnosis, the essence of which, as simulators testify, he is perfectly capable of deciphering in the protocol, has been a bitter disappointment.

However, another version of the same history is possible. As experimental facts concerning hypnosis lose their strength, as the false similarity between laboratories collapses, the obsessive fear of experimenters seeking, as always, to distinguish between simulation and "true hypnosis," each time discovering the formidable capacity of simulators, becomes more remarkable. The history of experimental hypnosis is a history that is played out "with" the simulators and is one that has accorded them a key role. It can be seen as a veritable race, one team claiming to have produced an effect "specifically" tied to hypnosis and another succeeding in producing a similar effect by appropriately motivated simulators. Each new protocol, each new requirement of proof has turned out to be susceptible of "creating" new behaviors, each time posing the same problem: Will the simulators be able to reproduce it? In the history of experimental hypnosis, this would translate: Should this behavior in turn be recognized as worthless, because it is determined by the fact that the experimental subject "knows" what is expected of him and by the fact that he *consents* to adopt the role asked of him? In a different history, one might have wondered: What new role would consenting to simulate render these subjects capable of?

We will return to this other history. In the history that was ours, the race called into question not only the existence of a behavior specific to hypnosis but also the latter's very existence. Hypnosis would only exist if behaviors betraying an "irrepressible submission," the "truly hypnotized" person not being "conscious" of simulating submission, really could be opposed to simulated, and thus voluntary, behaviors.

The unconscious, or the unknown, thus takes center stage. Bernheim, just like Delbœuf, had stressed the primordial characteristic of the consent of the hypnotized person and the fact that she plays or simulates the role proposed to her. However, for both of them, it was a question of properly hypnotic, that is to say unconscious, consent and simulation, even if the subject believed that she remained free, that she could willingly stop simulating. But, as Mikkel Borch-Jacobsen[1] stresses, if the rule of the game of hypnosis includes compliance being unconscious (in an evidently non-Freudian sense), no testimony, either of the hypnotist or the subject, will be able to prevail over doubt, and the only sure difference between a person under hypnosis and an actor is that "the actor plays his role before spectators who know that he is playing, whereas the hypnotized person (or the hysteric) plays hers before spectators who think that she does not know that she is playing,"[2] who think that she is not in control of her playing. And, it must be added, for whom the spectacle is interesting if and only if "she doesn't know that she is playing."

The fact that the consent of the subject to the role is liable to go as far as making her play the role of one "who acts without knowing it," if that is what the situation proposes, is a nightmare for all experimenters and not only for specialists in hypnosis. What is truly irrepressible? Who is the dupe of whom? What new experiment will be able to establish that the subjects really have, as this experiment supposes, been duped and thus are testifying without knowing it? It is the whole practice of experimentation in the area of psychology, as well as the data gathered by questionnaires, where the questioned subject may well conform to the role that the questionnaire attributes to him, that is called into question. Who knows how far consent may go? As for the "irrepressible" errors of reasoning that cognitive psychology puts into play, wouldn't they conceal, for example, that the subject knows perfectly well where it is that the experimenter expects him to make a mistake, and what's more, that the experimenter expects him not to realize it? Psychology could thus be viewed as a systematic producer of artifacts, these "facts" which seem to prove but are in fact incapable of doing so, for it is the experimental setting, including here the confidence and the hope of the experimenter, which is responsible for them. The laboratory would no longer be the place of proof but a stage on which the experimental subject plays her role of dupe or naïf in order to satisfy an experimenter for whom it is essential that she not realize that she is acting.

At the time of the First World War, the conclusion according to which traumatized soldiers, presenting certain of the symptoms that Charcot had reproduced under hypnosis, were in fact merely simulators was proven by the adoption of some brutal procedures (torpedoing [torpillage], the barbaric ancestor of electroshock) aiming to make them disgusted with this antipatriotic acting. The solution was effective, but evidently that doesn't prove much; moreover, it was of no importance; what counted was the transformation of a victim into a soldier able to go "back to the front." In the laboratory of experimental psychology, on the other hand, the question is critical. If hypnosis "doesn't exist," if it is all "merely simulation," then it's the whole economy of scientific proof in psychology that is in question. The ability to enter into a role that requires one to appear to be duped, and to produce all the confirmations that the scientist requires in this regard, leaves the threat of assimilation with the music-hall stage pure and simple looming over the laboratory setting.

Certainly, one could object that the limits of the assimilation in question are easy enough to explore. Teeth would not be pulled out on stage, and one would not proceed to remove a breast. Léon Chertok, who struggled all his life so that hypnosis would "oblige thinking," attached great importance to the possibility of reproducing under strict control certain traits that were part of what many magnetizers of the nineteenth century knew well, notably anaesthetic effects or the formation of "blisters" (subcutaneous emphysema) through the suggestion of burning. For him, such traits revealed that hypnosis makes visible a profound modification of the relationship between what we identify as "body" and "soul."[3] But, like others before, he collided with a noninterest that revealed the singularity of the experimental approach. The "fact" constituted by a woman submitting without flinching to a painful operation of dental surgery situates the scientist in the position of spectator. And not only does he not have the initiative but he also had first to enter with this woman into a relation that permitted him to test the capacity of the person to produce herself as insensitive to pain. Such a fact will thus be rejected as "anecdotal" as opposed to what is considered as "scientifically established."

In the laboratory, tests that could make a distinct difference between hypnotic behavior and simulated behavior (since it is impossible to ask simulators to pretend to feel nothing while their teeth are being pulled out) are no more valid than the miracles of Lourdes. The laboratory requires that the tests bear on *groups of individuals defined as anyones* [quelconques], that is to say, liable to authorize reproducible statistical correlations, so that these tests can be the object of quantitative evaluation. One will thus "reproduce" tolerance to pain in the laboratory by asking *all* the individuals in a statistical group, those who simulate and those considered as truly under hypnosis alike, to evaluate pain on a scale of intensity. But it will obviously concern a pain moderate enough for the adherence to role-play to suffice to explain that the subjects, simulators or not, can declare that they feel nothing.

We are at the crossroads. As we have just seen, the vulnerability of hypnosis to being reduced to simple role-play is linked to its laboratory setting. But one might nevertheless be seduced by the critical scope of this thesis: "It is nothing but role playing, a game which one cannot even claim is played by the subject unwittingly, is imposed on him in an irrepressible manner." Such a thesis, in its polemical dimension, seems to mark a decisive progress that leads directly back to the image of a science "battling against opinion." And in this case, the scope of the critical thesis is broad, as "opinion" here includes most of the practitioners themselves, even the skeptic Delbœuf, who could not keep himself from thinking that the submission of his subjects was unconscious, as well as the ensemble of therapists who could say: "Alright, but even so, we can testify to the fact that hypnosis has a certain efficacy!" The critic may answer that if adherence to a role demands that one acts "as if" one was feeling better, this efficacy is faked too. Besides, he might also propose that the patient may be "playing the role of a sick person" and is thus profiting simply from the opportunity that his therapist is offering him to change roles.

The ensemble of practitioners who thought they were implementing a transformation by putting into play a "beyond of the conscious subject" could thus seem to have been duped. Whether this beyond is characterized in psychological terms (Kubie) or psychodynamic terms (Freud) is translated as an awakening (Roustang) or a reframing (Melchior) is in this regard typical: the therapist's expectations induce the behavior that his patient will obligingly adopt.

However, it is possible to resist the seductions of critique. And, I would add, it is important to resist them. Indeed, to yield to them leads to putting the experimental laboratory back in command, even though this laboratory has been the site of a failure.

When a statement such as "it was believed . . . [now] we know . . ." arises from a true experimental laboratory what is "known" is generally much more important for the enunciators than the belief that they claim to be refuted, because the response of all those who this new knowledge will interest will be "so then, we might be able to . . ." Thus, at the beginning of the twentieth century, physicists were able to say: "Philosophers believed that atoms were indivisible, could not be observed, etc.; we know that they are liable to disintegrate and, if we cannot observe them directly, at least we are now able to count them." But philosophical atoms were only part of the setting of the success. What matters is the adventure that this success will make possible.

On the other hand, in our case, the fearsome capacity to simulate, which is the object of a knowledge that destroys belief, is not of interest as such to the critical experimenters. It is something that poses an obstacle to the economy of scientific proof, what destroys the authority of experimental fact, what marks the failure of purification operations. The subject adheres to the project of purification, he cooperates, he obligingly presents all the expected traits: it's a catastrophe.

The notion of role playing as such is thus not the object of a "we now know." Its intervention signals the possibility of a judgment that lumps everything together: whatever the subject does, it amounts to the same thing, "it's nothing but role playing," "it's nothing but simulation." What was problematic has been eliminated, absorbed in a more general category, so general that it poses no problem and can be forgotten after having done its duty of elimination.

It is important to resist such catch-all notions because they are what give to the adventure of scientific knowledge the appearance of a general and polemical progress, of an epic, even, marked by denials incessantly imposed by scientific truth on our "beliefs." The catch-all notion does not answer a question in the sense that the possibility of answering would mark a success. Certainly it responds, but in a general way, the prime meaning of which is to differentiate between "bad" questions, questions that interest non-scientists, and the answer whose primary effect is to disqualify this interest as unworthy of being taken into account.

Catch-all notions populate our sciences. It is easy to localize them. Every scientific definition is positive, but some of these definitions communicate with an exclamation, while others gain their importance because their formulation affirms the possibility of reduction: "It's only . . ."

No one will say that, in the mouth of Copernicus and his successors, the full import of the statement "the Earth is a planet that revolves around the Sun" was found in the statement "the Earth is only a planet." If "celestial mechanics" today still offers a singularity, it is in so far as it constitutes a fertile field of experimentation for all the new technical possibilities of calculation that permit the symbiosis between observation and mathematical physics to be reinforced. This was the case in the eighteenth century, with the question of the threats of instability of the trajectories of Saturn and Jupiter, and throughout the nineteenth century, with the question of resonances crowned by Poincaré's famous "three-body problem," and today more than ever with the chaos theory of dynamical systems. For astronomers, Earth is not "only a planet": they want to "know everything" about the trajectory of the bodies that turn around the sun, to construct their description in their full particularity. It is of the utmost importance to them to determine the stability of the trajectory of Earth, even if the consequences of its possible chaotic character are only expected to become manifest well after its disappearance, engulfed in the sun becoming a red giant.

On the other hand, Darwin's successors are torn between those for whom biological evolution is the start of an adventure and those who made selection a triumphant catch-all explanation. Altruism? It is merely the result of kin selection, which, moreover, characterizes ants as much as ourselves.

If "scientific progress" seems to cover the totality of the domains of our experiences and practices like a uniform blanket, it is because where the creation that obliges the scientist to think and to risk does not succeed, catch-all notions that authorize him to judge take over, picking up the baton and assuring an apparent continuity. And this appearance confirms and nourishes the polemical vocation attributed to scientific judgment: since this pronouncement "diminishes" what we "believed" to be important, it must indeed be scientific.

The operation of judgment announced by the "it is only . . ." is present in all those foundational operations where a science "gives itself" a methodologically purified "object." And it goes without saying: what is disqualified by this purification are defined as "philosophical" questions, and the fact that a science, once created, expels philosophers and their questions from the territory that it appropriates becomes part of the commonly admitted definition of scientific progress. On the other hand, the singularity of the history "of hypnosis," such as it begins with "its magnetic origins," is to render apparent the polemical character of the operations that condition the plausibility of the catch-all notions peopling it. As we have seen, the statement "this is nothing but role playing" draws its plausibility from what is excluded from the laboratory, the "anecdotal" traits that no one could be asked to simulate. But from the beginning, the condition of the verdict returned against "Mesmerian crises" "it is nothing but an effect of the imagination" was a redefinition of crisis that subjected it to a polemical test.

Mesmer and his followers saw themselves as scientists. For them, the magnetic fluid was inscribed in the lineage of other invisible entities, like Newtonian force or the magnetic force of minerals the existence of which is established by their effects. In this case, the existence of the fluid was proven by the crises and healings produced around Mesmer's baquet. But did these effects testify in a reliable manner to such a cause? That was the question of the commissioners appointed in 1784 by King Louis XVI and the Academy of Medicine. But the commission did not try to understand or even observe either the crises or the healings. It did not observe but invented tests to which the fluid, if it was to gain the status of "scientific reality," should resist. Thus, a complicit magnetizer magnetized a "talented subject" without informing him, pretended to magnetize another subject, or again, blindfolding the subject, magnetized a place on his body while announcing that another place was being magnetized. The subjects let themselves be duped and the fluid was thus experimentally disqualified: either it did not have the power to impose itself independently of what Delbœuf would have called the "consent" of the subject or this consent itself was liable to produce similar effects without the fluid. In place of the fluid, the commissioners invoked what we can recognize as a "catch-all" notion, imagination.

While Mesmer's fluid was exposed to the risk of being put to a practical test, to which it succumbed, the notion of imagination proposed by the commissioners of 1784 effectively conferred on them the power to disqualify without so much as having to take the risk of defining this imagination that they invoked and thus without having to produce a reliable witness for this definition. In other words, with regard to Mesmerian crises, the commissioners had implemented the steps that always inaugurate the creation of the space of experimentation. They "extracted" from its context what was expected to respond to their questions and submitted it to a setting eliminating everything that was judged parasitic or anecdotal. With the Mesmerian subjects duped, the scientists had the initiative with their questions. But they did not accomplish that which is the signal of experimental success, that which makes the difference between an experimenter and a judge. The phenomenon they staged did not thereby become more interesting, capable of inciting new questions, of creating new situations where the role conferred on the imagination would be put to the test and would earn new determinations. Judgment was passed, "it is only the imagination." Move on, there is nothing to see.

But the singularity of the history doesn't lie only in the visible character of the redefinition permitting the judgment that gives plausibility to the catch-all notion. It derives from the way in which the violent character of the transformation of power relations brought about by this redefinition was described and characterized. The magnetizers were not convinced,[4] they protested, stressing that the judgment of the commissioners concerned a mere caricature and that the question of the healing obtained around the baquet remained open. What is more, they pleaded, the testimony of the duped subjects was not reliable. They had not imagined that they could be duped, they wanted to satisfy the commissioners, they said they felt something even if it wasn't much. And if the commissioners invoked the power of the imagination to explain the effects felt "without the fluid," solely by the subject's "belief," why couldn't this power also be invoked to explain that the subjects had not witnessed the action of the fluid when they were unknowingly subjected to it? Imagining they had no reason to feel anything at all, why exclude the possibility that the power of their imagination may have thwarted this action? Didn't a young woman, who had been magnetized without knowing it, later explain that she had had to struggle to maintain the behavior demanded by the situation? She believed she was being interviewed for a job as a lady's maid.

In fact, the staging set by the commissioners is demonstrative only if a presupposition is accepted: if the fluid exists, its action must be general and irrepressible, like the action of gravitational force, for example. Just as gravitational force is indifferent to the difference between an accidental fall and a suicidal fall, the fluid must act indifferently, whatever the circumstances. Certainly, around the baquet, its effect could be amplified, concentrated, as Mesmer claimed. But if it exists, its action must be independent of what the subject "believes." Correlatively, if belief has an effect in the absence of the fluid, it is this belief that rightfully explains all that is attributed to the fluid. The magnetizers' counter-arguments made this presupposition contentious: if the action of the fluid is not independent of the circumstances, but implies what the commissioners called imagination, if "moral causes" are capable of amplifying or thwarting the "physical causes" (the fluid), the commissioners' demonstration no longer holds.

When there is experimental controversy, the central theme is always that of power and abuse, of reliable or extorted testimony: it is necessary that the experimental relation of force be recognized as corresponding to the effective submission of the phenomenon to the categories that this relation puts into play. The experimental setting can then be assimilated to a simple purification making this submission manifest and not to the fabrication of a phenomenon to which submission would *be produced by the relation of force*. In this case, there was no controversy: the commissioners were not looking for experimental success, they did not explore the possibility that the "fluid" acted in a way that wasn't one of physical force, indifferent to circumstances, nor that imagination itself might be also defined as a "force," to which we would belong, rather than as an attribute that belongs to us. These questions, and many others, were cultivated by the magnetizers' tradition. As for the commissioners, they had no other goal than to confer scientific authority on academic judgment against magnetism.

The protestation "but that's just a caricature" resounded again when hypnosis was defined, by Charcot and his colleagues, in opposition to the magnetizers.[5] What the new hypnotism defined as anecdotal, able to be eliminated so that the purified phenomenon would lend itself to scientific proof, is no longer a "crisis" but the rapport uniting the magnetizer and his subject, which makes them a veritable "couple," a magnetic rapport that had to be nourished and cultivated. In contrast hypnotists will use procedures of induction that break the couple up, demonstrating that they have nothing in common with their subjects and permitting them to adopt the posture of the true scientist who manipulates and shows, without being compromised, without figuring as partner in the show. Such procedures are brutal and authoritarian, even barbarous, some magnetizers protested, and they distort what they claim to reproduce. With such procedures the specialists in hypnotism will never attain what gives magnetism its interest, in particular the magnetic lucidity that makes the magnetized into a clairvoyant. They will only be able to produce caricatural effects, diminishing the person instead of lifting her up, analogous to circus freaks meant to impress the onlookers. And what is more, the type of efficacy privileged by their procedures designates neurotics as their privileged target.

Taking these criticisms seriously does not mean proposing a "return to magnetism." It concerns taking seriously the revolt of magnetizers faced with the radical transformation that the will to do science, to do respectable science, conforming to the canon of the experimental method, might make what is studied undergo. Because this method gives the scientist who elaborates hypotheses and questions the initiative, on the one hand, and demands submission from what must answer the questions, on the other, it was expected to eliminate what the culture of rapport assumes, the co-apprenticeship that produces a true couple, the magnetizer and his clairvoyant—most often, a woman. Hypnotism will define the hypnotist as interchangeable and name "induction" an ensemble of procedures without memory, making what is induced a fundamentally anonymous phenomenon, a phenomenon that can no longer be defined according to the coordinates of anything other than psychopathology (Charcot) or psychology (Bernheim).

The ensemble of resources with which the magnetized person could enter into contact is eliminated as illusory, a superstitious excess to what must be defined as an object of science. Correlatively, while magnetism was organized around a becoming, a metamorphosis of the person who becomes capable of what no "normal" person would have been able to accomplish, from now on the behavior of the hypnotized person will have to testify in a reliable way, but unwittingly, to the impotence that defines her. Indeed, what interests the hypnotist is the submission of his subject, what she is *incapable* of not executing when under hypnosis.

And here again, history repeats itself, but this time it takes as its victim the authors of the experimental setting. What Charcot's patients executed, Bernheim and Delbœuf denounced as nothing other than a "product of culture," the culture of Salpêtrière. With the theme of suggestibility, culture is no longer what is cultivated, but what, in an insidious manner, comes to live as a parasite on the hypnotic relation, to destroy the reliability of what it puts on stage. Until finally this new catch-all notion, "suggestibility," sounds the death knell of hypnotism itself.

The history associated with what we call hypnosis may indeed stutter, but it is nevertheless traversed by a remarkable continuity. Imagination, suggestibility, role playing could only claim to "explain" to the extent that what was expected to be explained was first reduced to the amnesic and anonymous conditions required by experimental proof.

While other "catch-all" notions may be linked to endeavors at conquest that are more or less rhetorical—from the scientific point of view, all of that can be reduced to . . ., those that have arisen throughout the history of "hypnosis" are inseparable from effective operations of redefinition. They thus offer the interest of illuminating the radical change in the imperative of proof when it intervenes in the so-called "human" sciences. The laboratory becomes the site of a "mobilization in the service of proof," implying, as does all mobilization, a breaking of the attachments constitutive of a person. You are no longer a peasant, a son, a father, you are a soldier and nothing in your past will get in the way of your availability to follow orders. You are no longer a patient who hopes, around Mesmer's baquet, you are no longer a clairvoyant of impressive gifts, you are no longer a talented subject, capable of having a tooth pulled without flinching, you are all "subjects" about whom a single question is posed: Does anything remain when nothing hinders your availability for the proof any longer? And the catch-all notion intervenes when the answer appears: nothing special.

Is it possible to escape mobilization in the service of proof? And at what price? Before approaching these questions, I must look at a very particular case, doubtless the only one where this mobilization provoked a true creation. The singularity of the Freudian creation of the unconscious, as I am going to present it, does not depend on Freud having made the choice, unlike the experimenters, of "listening to his patients." It depends rather on Freud listening to them with the same criteria of judgment as the experimenters: if what they recount cannot be mobilized for proof in some way or other, it is worth nothing.

## Notes

1 See Mikkel Borch-Jacobsen, "The Bernheim Effect," Ch.5 of *Making Minds and Madness: From Hysteria to Depression* (Cambridge: Cambridge University Press, 2009).

2 Ibid., p. 116.

3 See *Le Non-Savoir des psy* (Paris: Payot, 1979) (republished by Empêcheurs de penser en rond). Léon Chertok had chosen to ignore the "extra-lucid" traits associated with magnetic clairvoyance. He knew well the difficulties of the proof in the matter and preferred to concentrate on what, perhaps, would pose an acceptable problem for his contemporaries. "The boat has a big enough load" he said to those who reproached him about it.

4 See Chertok and Stengers, *A Critique of Psychoanalytic Reason*.

5 See Bertrand Méheust, *Le Défi du magnétisme*, volume I of *Somnambulisme et Médiumnité*, pp. 533–49.

# 4

# *Freud's Coup de Force*

Unlike experimenters, most contemporary therapists practicing hypnosis don't accord any particular importance to the question of simulation nor, more particularly, to the question of knowing if their patients know whether they are simulating or not. What counts for them are the practical effects of the relation, and if their patients have the talent to simulate getting better, it's because they have consented to get better, which itself makes them feel better. But this pragmatic humor is inseparable from a history in which for most the fact of practicing hypnosis meant freeing oneself from the psychoanalytic reference to truth. That, in itself, marks the singularity of Freud's role in the history that concerns us, the veritable "stroke of genius" of which he is the author. Mobilized in the service of the proof, experimental subjects remain, or become, indifferently, "anyone," whereas Freud's patients "mobilized themselves." Living proofs of analysis, many will become its vectors, for, as Lacan justly remarked in 1964, "psychoanalysis presently has nothing more certain to assert in its favor than the production of psychoanalysts."[1] Today, in a culture still dominated by the grand theme of psychoanalytic truth triumphing over the seductions of fiction, the imaginary, and of opinion, those who dare to practice hypnosis must first be defined as "demobilized," as having found, at their own risk, the path of a certain indifference with regard to the polemical passions of truth.

We know that Freud never stopped affirming that his technique had nothing in common with hypnosis (or with suggestion). This affirmation is only convincing for those who accept the image of hypnosis that Freud constructed, for Freud never discussed the interpretation that a Delbœuf, and with him every connoisseur in the matter of hypnosis, would have been able to propose of his "facts." He limited himself to making a contrast between his technique and suggestive hypnosis à la Bernheim, which privileges a direct and verifiable relationship between the content of the administered suggestion and the response of the subject. He always affirmed that he never imposed a memory or an interpretation on his patients (à la Bernheim). Delbœuf would not have doubted that for an instant, but would have remarked that, strictly speaking, that proved nothing.

When he proposed his self-serving vision of it, good enough for those who trusted him, Freud was presupposing and betting on the abandoning of the disappointing and deceptive hypnosis, which was in the process of happening at that time, and thus on the disappearance of those who would have been able to contradict him. However, if the threat that hypnotic compliance might bring to bear on analytic facts is never evoked in his texts, perhaps it is only because it is all the more present. For it is possible to affirm that the definition of the Freudian unconscious is indeed centered around the necessity of warding off this threat and that it is even from this that it draws its singular originality.

Numerous are the psychoanalytic thinkers who have shown in page after page that what Freud's unconscious has in common with the discoveries of modern science is its radical novelty, sharply cutting through all prior, philosophical or scientific, conceptions of the unconscious. As such, this unconscious thus signals the rupture with the pre-analytic past and in particular with the unconscious of sleepwalkers. And certainly, Freud's unconscious is not at all comparable to the "unconscious self" evoked by Delbœuf, for example. But it is precisely this difference, this "it has nothing to do with. . ." that is interesting, and not in the sense of a Freudian denial but in the sense that the problems are effectively different. While Delbœuf's unconscious designates the compliance of the hypnotized subject who, unknowingly, actively fulfils suggestions to which he seems to be subject, the Freudian unconscious is *tailor-made to protect analytic technique from the suspicion that reference to this compliance could have a bearing on analytic facts*. It is indeed first through the resistance that it opposes to pertinent interpretations that the existence of the Freudian unconscious imposes itself in the course of the analytic process. Freud will thus take as witness not the adherence of his patients to his interpretations but the ensemble of effects of this resistance, which feed and relaunch the analytic work.

The caricature of hypnosis that Freud used in his argument is thus a stooge, an integral part of the "legend" of psychoanalysis. And perhaps we must remember here that legend first signifies "what must be said": what Freud himself wrote so it would be repeated, so, also, that other questions would not be posed. The power of this legend can be measured by the scandal produced by so-called revisionist historians of psychoanalysis, even though they limited themselves to doing their job, to asking the archives what happened.[2] Indeed prior to this scandal the history of psychoanalysis was saturated with expressions that mark the narrative of scientific discoveries: "Freud understood that . . .," "Freud recognized that . . .," and especially "Freud, at this time (when he was practicing hypnosis), did not yet know that . . . ."

The scandal is one of a kind. Of course, the history of the sciences is, as a whole, saturated with legends that historians of sciences have learned to "undo," most often to the benefit of richer and more subtle histories. But their work is not supposed to inspire the slightest doubt in scientists about the validity of what they have inherited. Even the narrative of Newton the alchemist only worried physicists with regard to opinion, which "might believe" that the success of Newton somehow legitimizes the tradition of alchemy. On the other hand, the investigations of the historians of Freud raise very thorny questions. Specifically, they show that at the time when he considered hypnosis a reliable instrument of investigation,[3] capable of bringing up forgotten memories, Freud indeed fell into the traps that Delbœuf had foreseen.[4] It would even be necessary to say that he went along with this because, as Mikkel Borch-Jacobsen has mercilessly shown,[5] the interpretative hypotheses that he formed, far from being inspired by his patients, were instead confirmed by the material his patients furnished after they had been formed.

Freud recognized this in veiled terms when, in 1914, he remembered with nostalgia the evocation of memories under hypnosis that gave "the impression of an experiment carried out in the laboratory,"[6] but he never admitted it. Not even at the moment when the conviction that he had been misled by his patients imposed itself on him: at the "decisive" moment when Freud becomes "our" Freud, the theorist of the unconscious, heroically abandoning the theory of seduction that his patients had, nonetheless, amply corroborated.

A terrible problem for Freud: How to disqualify the numerous testimonies he had obtained confirming this theory, without, at the same time, recognizing that he had guided the testimony of his patients, that is to say that they had obligingly consented to "testify" in favor of their doctor's theory? Considered from this point of view, the "discovery" of the unconscious becomes a true stroke of genius. Because the unconscious permits Freud to affirm both that these memories and the theory that they verified were indeed "false," and that, nevertheless, *they owed nothing* to his own insistence, his own interest, his own questions, in short, to his suggestive influence. His only error had been to confuse phantasies with reliable memories, designating real events. All his life, Freud denied that his patients could have "consented" to corroborate his theories. He chose to make the testimonies he had obtained into "screen" memories, thereby proving not only the connection between repression and the unconscious but also the limits of the hypnotic technique. This technique is effectively *retroactively* defined by the (bad) possibility of a short-circuiting of resistances, giving free rein to phantasies. And there you have it: before having "discovered" the unconscious, Freud could only misunderstand its power, could only ignore the question of resistances and the possibility of "screen memories." Analytic technique, which addresses the unconscious and is equipped to confront resistance, will produce reliable witnesses.

Must we henceforth assimilate this episode, which supporters of the decisive rupture with hypnosis celebrate as the correlated creation of "psychoanalysis" and the "psychic reality" that psychoanalysis addresses, to a story of deliberate imposture? Before turning Freud into a simple "trickster," perhaps it is necessary to give full weight to the real "scientific drama" that is outlined here. For, according to this "revisionist" version, it is not the possibility of a correctable "error" or of an artifact it was possible to eliminate that Freud refused to take into account. What was in question was much more serious, for it indeed concerned the very principle of the establishment of facts in psychotherapy. What Freud refused to envisage, or confront, was the destruction of any hope of making therapy and science converge. Indeed, the reference to a subject capable, in some circumstances, of unknowingly mobilizing extraordinary resources in order to "partner" in the suggested game, and of doing so without being forced in any way, *without there even being any need to explicitly ask for it,* transforms this convergence into a trivial success, that any therapist can obtain in regard to any theory and which thus proves nothing.

Witnesses who, with intense feelings of truth, produce a fiction that seems constructed to satisfy the criteria of proof are a nightmare for any tribunal to deal with. Having to accept his possible inability to distinguish between a consented fiction and an interpretation giving access to the truth was a nightmare for Freud.

That Freud refused this possibility and used all the means at his disposal to hide its insistence, from caricature to the definition of an unconscious characterized by its radical "non-compliance," detectable not by adherence but by resistance to pertinent interpretations, indeed constitutes a "stroke of genius," but in the strategic sense of the term: it is also a question of a bid for power.

The great singularity of analytic technique is thus not to have claimed to establish direct communication between the mode of intervention and the formulation of theory, nor to have claimed to have based theory on clinical experience. These two aspects are present in Freud's work, but they do not singularize him in relation to Charcot, Bernheim and an innumerable cohort of psychiatrists or cranks happily forgotten. The great singularity of psychoanalysis comes from Freud giving himself the means to escape the fate of everyone found to have based their claims on the pretenses of compliant testimony. He constructed a "psychic reality" that created its relation with science at the crucial point where what is at stake is resistance to the accusation "this is only a fiction." Freud's acknowledged literary talent doubtless also derives from this, at least in part. His case studies are written in the manner of investigations, where the skeptical author-investigator incessantly poses the question of the validity of his interpretations to show, progressively, why he is forced—and the reader with him—to accept them. This talent for setting the stage is not a mere add-on for the creator of psychoanalysis. The investigation in question reproduces the tests of a scientific controversy, one in which possibility and risk are integral parts of the construction of "psychic reality." The mode of existence of the latter integrates the possibility, for Freud, of forcing the skeptic to bow down before the facts.

However, prevailing in a scientific controversy is not reducible to the staging of a powerfully convincing construction. A controversy implies that the skeptics who are finally forced to bow down are competent colleagues, not objectors invented by the author strictly to fit the mode of construction of his psychic reality. The least one can say is that none of the objectors imagined by Freud has the competence of a Delbœuf, none will happen to recall that a compliant patient is evidently capable of presenting the type of resistance expected by Freud, that is to say, of testifying with talent for the difference between conflictual effects of a pertinent interpretation and a simple adherence that proves nothing.

But the foundation of a science doesn't answer the criteria of intersubjective morality, and as contestable as it may be, the idea that psychic reality testifies in a reliable manner precisely when the patient refuses to adhere, that is to say "resists," could have become one of those "masterstrokes" which are always scandalous for their contemporaries but are later celebrated even in their amorality by the inheritors of the field thereby founded. When the "masterstroke" works, even the bad faith that presided over its invention becomes part of the "genius" of the inventor.

Freud's "coup de force" thus did not a priori doom his ambition to create a "scientific technique," having as guarantor a "reality" conferring on his interpretations the capacity to resist the accusation of being only interpretive fictions. But for this to be successful, it was necessary that the validity of this reality be confirmed by a practical difference, by the privileged efficacy of the technique, which could be interpreted by competent colleagues, still skeptics but now interested, as proof that the analytic technique had indeed found reliable witnesses. Whatever the revisions of the legend might be, Freud's coup de force would thus have been able to enter *as such* into the prestigious succession of scientific successes. And in that case, his colleagues would have been forced to recognize that analysis indeed accomplished technically what was demanded in vain from hypnosis: success in conferring one "secular" truth on the exotic multiplicity of more or less efficacious modes of healing associated with therapeutic techniques referring to ancestors, sorcerers, spirits, charms, divinities. Freud's unconscious would exist because without it this success would be inexplicable.

The history of psychoanalysis is, as we know, saturated with dissent that Freud himself interpreted as the resistance of [some of] his followers due to a badly conducted or prematurely concluded analysis, and this interpretation itself constituted the admission that a "successful" analysis is at the very least an event that is rare and in any case difficult to confirm. Besides, revisionist historians have stacked up the cases where Freud spoke of a cure, while the unhappy patient, duly analyzed, continued his sorrowful career as a sick person. But in 1937, in *Analysis Terminable and Interminable*, Freud himself finally recognized explicitly that from the point of view of its efficacy, analytic practice leaves much to be desired. Does he admit to being beaten, that everything has to be started "all over again"? No, for he no longer speaks as a pioneer needing to convince others, but as the leader of a school addressing the convinced and wishing them good luck in their "impossible profession": the theory is good, he affirms, and the proof is that it allows the practical failure of the technique to be explained. Psychoanalysis comes up against resistances that are quantifiably too strong; it cannot be asked to prevail over them even though it can interpret them.

This last "coup" of Freud's was enormously successful. It allowed the ink to flow, celebrating the heroism of the psychoanalyst, who confines himself to what the truth of human suffering obliged him, who does not let himself yield to the temptation of techniques that are possibly more effective, but only in the "short term," of course. It is a question of resisting regression, toward hypnosis, for example, which in the first place manifests the incapacity to tolerate the constraints that define the greatness of the "Freudian rupture."[7]

It would be tempting to analyze the history of psychoanalysis in hypnotic terms and notably this celebration of the 1937 article as a remarkable manifestation of Freud's lucidity, of the manner in which he does not hesitate to expose the limits of his technique. One could also analyze in these terms a long series of texts that appear critical of psychoanalysis but, very curiously, use psychoanalytic theory itself to account for its weaknesses or to diagnose the reasons why Freud ignored such or such an aspect of a situation. There is much material here that is a marvelous illustration of the talent, attributable to hypnotized persons, of actively transforming what should turn them away from a theory into its confirmation, in this case the realization of the manner in which this theory confers a theoretical universality on traits that are now legible in historically, culturally, and socially situated terms.[8]

As tempting as it might be, such an analysis would confer on hypnosis the characteristic of stable reference that it precisely does not have. It would be better to proceed by contrasting the two types of history, two histories whose common trait is to be "without progress" even though they are haunted by the model of scientific progress. While this model stutters throughout the history of hypnosis, it can be said that psychoanalysis captured and reproduced one of its essential traits by entirely different means: the stable assignation of roles. From this point of view, Freud's "coup" is indeed a "masterstroke": like the scientist in the laboratory, the analyst occupies a position that the person he is dealing with cannot contest. More precisely, if he allowed himself to be called into question, he would have to work through this error on another analyst's couch. It is not surprising that with the exception of Ferenczi, who took the extraordinary risk of mutual analysis, none of the psychoanalytic dissidents questioned this aspect of the "setting." Jacques Lacan even accentuated it with the invention of "short consultations" where the analyst is no longer even constrained by a contract bearing on time but may unilaterally (an "act," Lacanians say) interrupt the consultation.

Correlatively, while the history of hypnosis produced lucid interpreters, caught up in controversies, the radical nature of which went so far as to question the very existence of what brought them together, psychoanalysis has produced a mode of bringing together that may evoke the consensus of scientists after the successful closure of a controversy, aside from the fact that here the closure passes via a triple disqualification. What poses an obstacle to the cure has been disqualified, reduced to the simple question of quantitatively insurmountable resistances. The skeptics, whose "resistance" is explained by the refusal of the narcissistic wound of the unconscious, have been disqualified. And little by little, the dissidents are disqualified, the object of their dissidence being interpreted as a "pre-analytic" regression, of which their very dissidence itself is, moreover, the symptom.

This mode of closure confers on the "truth of analysis" some of the traits that could be associated with an initiation procedure. It is indeed legitimate that a form of adherence constitutes a condition of possibility for access to a truth of the initiatory kind. However, the question that then arises is the one which is doubtless posed for any initiation procedure, the question of safeguards created against modes of blind adherence. However diverse the tests put in place by each one of these procedures might be, these tests go back to a truth that is non-modern in Foucault's sense, a truth that cannot be attained without the modification of the one who attains it. That is doubtless why none of these procedures would admit the type of judgment in terms of which psychoanalysis presents itself: a general, public judgment, enunciated with the neutral authority appropriate to those who represent the power of rationality and progress—we are sorry, but we have the right to announce to you all, as a fact and not as part of the initiation process, that you are living an illusion. It can thus be affirmed that psychoanalysis has suppressed all the safeguards at once: the tests of controversy proper to the sciences that allow scientists not to have any specific lucidity, and the tests appropriate to the initiation procedures which do have to verify such lucidity.

Hence, it is not impossible technically to bring this double suppression into closer proximity with what defines a sect. Unfortunately, the term "sect" is today associated with an insult, even the threat of being made illegal, not with the technical analysis of its own type of efficacy. Yet it is this efficacy, which gives psychoanalysis a type of consistency[9] that neither magnetism nor hypnotism had ever gained, that matters. What is indeed striking is the correlated setting in motion of heavily charged themes, about which it can be said that they possess us or that we are their prey.[10] None of them are new as such, what makes for an event here being a capture which opens them up to one another and confers on them significations that derive, first of all, from the way they refer to one another. I will cite, in a non-exhaustive manner and in no particular order, the technique of confession, pinpointed by Michel Foucault, the fascination for an admission of truth that liberates, the truth as detaching from illusions, the search for a universal beyond the attachments that blind and divide, the critique of fetishes, original sin and redemption, the fatal weight of a past defined by a forgotten crime, the obsessive fear of what manipulates us without our knowledge, the white man's burden, the disenchantment of the world. None of these themes explains the others, all communicate without any one being their common root. It is a question of what Deleuze and Guattari call a "machine."

It pertains to the very extraordinary Lacanian machine, which took up the baton from the Freudian machine, to have successfully dared to risk an explanation of the mode of functioning of the efficacy of analysis. Because the set of themes that I have just cited would make a Lacanian snigger. He would read there, without any problem, the imaginary that defines the psychoanalytic subject and signals that psychoanalysis belongs to the Western world but does not define its truth. Lacan's master stroke has indeed been to dare to cast off the moorings that attached Freudian psychoanalysis to a reference of the scientific type. With the theme of the analyst as "subject supposed to know," whose fading will mean the end of analysis, and with that of the analysand as "split subject," who must first of all be cured of his desire to be cured, to be mended, to be saved, the Lacanian machine succeeded in capturing and even setting in motion the most anguishing themes of Western thought, those that speak to the vanity of all positive knowledge.[11] The truth of analysis transforms into so many of its component elements the theology of the absent God, the infinite wandering of a man guilty of a rupture that is, however, the basis of his freedom, the illusions of a subject captivated by what permits him to forget his finitude, the efficacy beyond reason of what frees him (the efficacy of the analytic act is no more deserved than is grace in Saint Augustine).

What springs up where a machine mobilizes and sets in motion is the power of an abstraction, whereby judgment is passed concerning what is mobilized, and which is henceforth reduced to its role in the machinic functioning. This does not in itself signify a denunciation: machines, as Deleuze and Guattari define them, populate the world. But it does signify the redundant character of the history that sets the stage for the triumph of this abstraction. Thus, it would be redundant to recount a history that, from Mesmer to Lacan, would go from the power of a natural fluid, revealing the "animal" nature of humans, to that of infinite debt, testifying to the solitude of the subject prey to a desire that separates him from all "nature" (a theme that Deleuze and Guattari, after Nietzsche, associated with the "priest," the one who separates those he deals with from their power to act and to feel). But other apparently more perplexed narrations equally mark their belonging to machinic functioning. This is notably the case when Octave Mannoni writes: "We eliminate the devil, convulsions remain. We eliminate relics, Mesmer's 'magnetized' remain. We eliminate the baquet, we have hypnosis and the 'rapport'. We eliminate hypnosis, the transference remains [. . .]. We must resign ourselves to treating the capacity to transfer as a 'trait of our nature.'"[12] The resignation which closes this history of successive eliminations does not, as it would seem, exude modest lucidity as to a history that would not have kept its promises. Quite the contrary, for the final "remnant," the product of the final abstraction, is a "trait of our nature," which, as such, may claim to transcend history and cultures. *Nec plus ultra*.

Machines render discourses of the epistemological type redundant, for those discourses limit themselves to conferring on the abstraction that signals the setting in motion of a machine a status that justifies its power and ratifies its judgments. Machines, in contrast, call for an evaluation of a pragmatic type as to their effects. Pragmatically, the assurance of the judgments of analysis cannot help in prolonging the permanent stuttering associated with the history of "hypnosis." This whole history incessantly cast doubt on the very existence of what motivates it, while every psychoanalyst, from the first Freudian to the most subtle Lacanian, testifies to the "existence of the unconscious" as the "cause" of psychoanalysis.

To try to prolong the history of hypnosis today is to make a choice as to the manner of prolonging a heritage wherein the disparate components that the Freudian machine effectively succeeded in articulating in fact cohabit. In this case, it is to make the choice of attaching oneself to what forced the stuttering, the permanent disarticulation. And this choice implies taking seriously the scientific requirements that the specialists in hypnosis tried to satisfy and that—differing in this from most specialists in the human sciences—they were constrained to take into account in their singular dimension: requirements not designating the submission to an at last rational, objective, methodological approach, but the risky attempt to constitute a phenomenon as a reliable witness to the way in which it must be interpreted. Because it is the repeated failure of this attempt that has made the history of hypnosis stutter. And if this failure is taken seriously, it might oblige thinking against the "methods" that elsewhere take up the baton when witnesses betray and that confer on scientific progress the allure of a uniform conquest destined to redefine all of human knowledge. But to do this, one last temptation must be avoided, that of concluding too rapidly that "apart from reliable witnesses, there is no salvation."

# Notes

1. Cited in François Roustang, *The Lacanian Delusion*, trans. Greg Sims (New York and Oxford: Oxford University Press, 1990), p. 17.

2. The existence of these works called "revisionist" was revealed to the French public with the "great" controversy on the adjournment of the Freud exposition in Washington and the protest petition that ensued. In view of the accusations brought against the "revisionists," one may conclude that it is, here again, the question of "consent," but this time to the Freudian discovery, which is at the center of the affair. The revisionists, to the extent that they do not consent, must be disqualified, as must be excluded from the laboratory of experimental psychology anyone who, instead of consenting to serve the science, discusses what is wanted of him. See Mikkel Borch-Jacobsen, "Une visite aux Archives Freud," in *Folies à plusieurs*.

3. We know that Freud rather quickly abandoned open induction, with the declaration of hypnosis, and stopped putting his patients to sleep in order to adopt a technique that exposed him much less, the "*Druckmethode*": pressing his hands hard against his patients' foreheads, he declared to them that they were going to remember. But once again such "new" procedures—including, it is necessary to make clear, the technique of so-called free association—are procedures derived from hypnosis.

4. Delbœuf, "M. Liégeois et les suggestions criminelles," in *Le Sommeil et les Rêves et autres textes*,p. 369: "It is necessary to take into account the exact appreciation of the circumstances by the *unconscious ego* of the hypnotized subjects, for it concerns in these conditions a highly awakened psychic activity, even though the subject seems to be plunged in a profound lethargy. This unconscious personality played remarkably bad tricks on observers looking for proofs of mental suggestion, of lucidity or other phenomena or faculties still highly controverted, and about which experimentation, by the fact of this unconscious activity, is of the most delicate and the most frustrating."

5. Borch-Jacobsen "Neurotica," in *Making Minds and Madness*.

6. Sigmund Freud, "Remembering, Repeating, and Working-through," *The Standard Edition* 12 (1950, first published 1914): p. 152.

7. See notably the theme of the "hypnotic symptom" in Elizabeth Roudinesco, *La Bataille de cent ans: histoire de la psychanalyse en France: 1885–1939*, vol. I (Paris: Seuil, 1986), especially pp. 168–9.

8   Transformations of analytic theory are often presented as so much progress, but, from a scientific point view, it is a question of pseudo-progress. In the normal histories of scientific progress, it is the very dynamic of the research that permits the recognition of weaknesses in the starting point, while, in this case, it is the politico-cultural events that play a decisive role, for example decolonization, feminism, the movement for homosexual liberation, and who knows, perhaps tomorrow the marketing of psychotropic drugs . . . Here again, it is necessary to read Mikkel Borch-Jacobsen, "A Portrait of the Psychoanalyst as Chameleon," in *Making Minds and Madness*.

9   In the sense of Gilles Deleuze and Félix Guattari in *A Thousand Plateaus*, trans. Brian Massumi (Minnesota, MN: University of Minnesota Press, 1987), where "consistency" designates the manner in which heterogeneous groupings hold together as heterogeneous.

10  These thematics are not, in this sense, "ideas" (history of ideas), but designate rather "assemblages" (agencements); the idea has as a correlate the subject who "has" the idea, in general, whether he thinks about it or not, while the assemblage, as Deleuze and Guattari define it, does not exist "in general," but always in the mode of here and now, in the mode of capture. It is thus apt to being described by a verb in the infinitive because it is the capture designated by the verb that brings into existence what the verb articulates. The assemblage is the real subject of the utterances, which describe its effects on the ensemble of what finds itself "captured."

> The minimum real unit is not the word, the idea, the concept, or the signifier, but the *assemblage*. It is always an assemblage which produces utterances. Utterances do not have as their cause a subject which would act as a subject of enunciation, any more than they are related to subjects as subjects of the utterance. The utterance is the product of an assemblage—which is always collective, which brings into play within us and outside us populations, multiplicities, territories, becomings, affects, events. The proper name does not designate a subject, but something which happens, at least between two terms which are not subjects but agents, elements.

Gilles Deleuze and Claire Parnet, *Dialogues* (London: Athlone, 1987), p. 87.

11  According to Deleuze and Guattari, what assures the consistency of a machinic assemblage is always its most "deterriorialized" component, the one that is most susceptible of passing into and relaying toward other assemblages in which new components participate. In our case, it seems to me that this component is none other than the one that permitted Freud to affirm the invulnerability of psychoanalysis in relation to the disappointments of hypnosis: the node between truth and negation ("what *is not* a fiction, an object of adherence"). It is this node that transforms all the positive components of a life into what, as the case may be, will have to come to feed the analysis, and it is this node that Lacan realized could "hold" without the slightest positive counterpart (without "therefore, it was that!"). A possible confirmation of this hypothesis could be the ravages inflicted by the Lacanian machine on the Jesuit order. The Jesuits are indeed the Pope's "army," first mobilized against Protestants, then against non-believers, and their membership in this army imposes their ironic

indifference to all positive questioning of the Roman Catholic Church and its dogmas. The "spiritual exercises" destined to make them invulnerable to the demands of reason that nourish the heresies, skepticisms, "naïve beliefs" of free thinkers, have on the other hand, made them vulnerable to a capture operation by a machine that feeds on the vanity of all reason.

12  Octave Mannoni, *Un commencement qui n'en finit pas: Transfert, interprétation, théorie* (Paris: Seuil, 1980), p. 50.

# 5

# *It's Only an Artifact?*

The great hope associated with hypnosis designated it as that which, suitably purified and performed, would constitute the finally scientific access to a multiplicity of more or less disturbing manifestations that seem to reveal supernatural powers, whether these manifestations belong to therapeutic, magical, or religious registers, whether they concern wild ecstasies, possessions, or ritual trances. And so, as it is the canonic hope of our histories of progress, the purified phenomenon might give way to the setting up of "scientific techniques" authorized by a finally scientific reading of what they address. Thus, it can be said that the ensemble of metallurgical techniques of today address what research laboratories learned to read in terms of chemical compounds, making ancient arts, which were successful without anyone knowing why, tumble into a prescientific past.

This hope has been disappointed and deceived: the different stagings of hypnosis have produced divergent data, inseparable from their particular mode of production. This implies that the staging does not give what is interrogated the opportunity to manifest itself in an at last readable mode, but that this staging is the author, or in any case coauthor, of what is read. From the point of view of experimenters, the facts are "badly fabricated." They are incapable of testifying in a reliable manner to a unique reality beyond the multiplicity of its manifestations. They are incapable of demonstrating that this multiplicity is only a throwback to the multiplicity of cultural, mystical, superstitious interpretations that obscured their common explanation. In short, these facts are not, in the end, different from those multiple manifestations of which they claimed to produce a purified version. They are, experimenters will say, "artifacts."

It will be understood that this failure makes it entirely possible to devise a general operation of critically questioning a science like psychology, the most diverse categories of which—from motivation to the psyche itself, passing through information-processing or modified states of consciousness—all participate in the same history, which seeks to constitute a human universal beyond the multiplicity of practices and interpretations. The scope of questions provoked by hypnotic compliance, or by the talent of those who lend themselves to the suggested role-playing, cannot be limited at will. In fact, they have transformed the specialists in hypnosis into perspicacious critics of what their psychologist colleagues accept as "data," whether produced in the laboratory or collected via questionnaires.[1]

And this perspicacity offers the most singular contrast with the relationship that theorists of psychology maintain with "facts."[2] I will limit myself on this point to an example raised by Ian Hacking, who underlined the very different reactions of two researchers in psychology, Daniel Dennett and Martin Orne, confronted with the epidemic of "multiple personalities" that -raged relatively recently in the United States.[3]

Daniel Dennett is a theorist of neuro-cognition, for whom the possibility that consciousness may be explained in terms of neuronal treatment of information belongs to the very order of scientific progress. And, in *Consciousness Explained*,[4] he speaks, in regard to multiple personalities, of a "terrible experiment undertaken by nature," which translates well the extraordinarily precious character (for him) of such a case. For "nature" (in this case, Dennett accepts, a horrifying history of sexual abuse) is here charged with producing what the researcher cannot, for ethical reasons, envisage creating in the laboratory. And what it produced is a veritable godsend for Dennett, a veritable illustration of his theory: in this privileged case the same psyche exhibits distinct, differentiated, split modes of information processing.

On the other hand, for Martin Orne, a specialist in experimental hypnosis, "nature" has nothing to do with this production. Nature did not create any reliable witness for anything. The "revelation" of multiple personalities harbored by the patient as well as the terrible trauma that, as the diverse personalities testify, is responsible for the dissociation, proceed, explicitly or not, from a treatment of the hypnogenic type. For Orne, since both the patient and the doctor are looking for a hidden truth, it is useless to protest (as did Freud) that the testimony has not been forced. The patient knows where he is, which kind of therapist he is dealing with, and thus what diagnosis has been made. That is amply sufficient for him to be able to "fill in the blanks," take on the anticipated role and remember traumatic episodes that determined the dissociation.

Today, the fact that multiple personalities constitute an artifact of the treatment that was supposed to allow them to express themselves is generally recognized. Legal cases brought by accused parents, then by patients, have provoked a panic in the ranks of those mobilized by a therapists' crusade that created an ever-growing crowd of "multiples" as so many witnesses of ever more horrible abuse. But it was at the height of the epidemic that Martin Orne had the courage to refer to what he had learned from hypnosis: the terrible ease with which hypnosis, when it deals with "gifted" subjects, can incite (false) witnesses who are the most convinced and thus the most convincing, adhering with all their strength and with an exceptional talent to a fabricated truth.[5]

But is it necessary to admit that the courage and lucidity of Martin Orne constitute the true end of the history of hypnosis, of what it has to teach us? Such a conclusion may be tempting. It would indeed signify a new conquest of the critical mind, something we have rather automatically tended to consider as a good in itself: let everything that cannot resist the acid of critique disappear and dissolve! But we must not be misled, for from this critique nothing else would be born but a rather monotonous series of denunciations: this is an artifact, so is that, and so is that as well . . .

In his *Making Minds and Madness*, which I have cited on numerous occasions, Mikkel Borch-Jacobsen presents "fragments" of what, he states, should be developed as "a theory of generalized artifact," beyond the opposition between "real" and fictive": all realities are "constructed by two, by several, by many," in the manner of the "elements of an ecosystem [which] accommodate each other to create a particular plant, animal, or virus. Faced with such entities, the question is no longer whether it is real or not, but only whether it propagates or not, and how" (p. 108). However, this program is ambiguous. For it may oscillate between two orientations, either critical irony (in which Borch-Jacobsen engages with enormous talent) or taking seriously the analogy with biology, and more precisely, with etho-ecology. And in this case, the project of a "generalized theory" appears highly adventurous because there is no example of an analogous achievement in etho-ecology: each situation creates an articulation of distinct problems, without authorizing the economical type of generalization that the notion of a theory announces. More precisely, there exists in biology a theory that claims to unify the ensemble of questions articulated around living beings and which is effectively centred on the question, "Does it propagate or not?" This is the theory of selection (the living being as a vehicle of its "selfish genes"). And in this case, Borch-Jacobsen's theory already has a name: it's the theory of "*memes*," which self-replicate like genes, but infect our brains. Here also, the only question to ask is does a *meme* (an idea, a melody, an "evidence," a manner of judging, a manner of critiquing, etc.) propagate or not?[6] *Memes* are to cultural evolution what genes are to biological evolution: all-terrain notions that make every situation "amount to the same"; that is to say, that guarantee the pertinence of a unique explanatory hold around which everything else will have to be organized.

Certainly, Mikkel Borch-Jacobsen adds to the question "does it propagate?" the question "and how?" But if this "how" is taken seriously, fully deployed, it is the ambition of theory that will be called into question, because the "how" will not authorize any economizing. As is the case with Deleuze and Guattari's "machines," it will, in each case, be necessary to learn how to map the mode of setting in motion of heterogeneous resources. Back to the empirical story of captured events.

Moreover, the term "artifact" reproduces this hesitation. Either it designates the ensemble of "facts" implying the arts of humans. It is certainly possible to say, in a first approximation, that a catastrophic earthquake and a tidal wave are not artifacts. However, everything that humans take an active interest in, from experimental facts to emotions,[7] is an artifact, and the idea of a theory bringing together the ensemble of these "facts" is, to say the least, grandiose. Or else, artifact designates the negative counterpart of the notion of reliable witness, which belongs to experimental practices, and so one will then say, "this is only an artifact." But in this case, we come back to the type of monotonous hold that genes and *memes* permit: a pseudo-experimental hold the primary interest of which is to contribute a continuous stream of canonical "you believe that . . ., but from a finally scientific viewpoint, it is only . . ." statements of scientific progress.

Taken in its pejorative sense, the notion of artifact functions in a manner that is somewhat analogous to that of "placebo":[8] as the "other" of the success being sought—reliable witness in one case, molecule whose therapeutic efficacy is superior to that of the placebo in the other. It does happen that persons of goodwill affirm the necessity of a "theory" of the action of placebos, which theory would, they hope, permit a rational foundation for the efficacy of therapies "without drugs," including those that require objects whose efficacy is not "scientifically demonstrable." The problem is that it is difficult to conceptualize what a "theory of the placebo" would be, because theory can only be conceptualized when a phenomenon is cultivated "for itself," when study allows it to deploy what it is capable of producing in different circumstances. The very definition of the placebo is that it cannot be studied in this way. The placebo is the residue, the curative effect, that survives in the sanitized setting that clinical trials require when the patient has absorbed a product devoid of any particular effect. The placebo effect thus has no other status than that of background noise, what cannot be eliminated in spite of the strict purification imposed by the clinical trial's imperatives of statistical reproducibility. Its mode of existence is determined by the judgment that defines as parasites all manners of healing that do not authenticate the tested molecule as playing a part in the healing process. In the same way, the artifact has the status of failure in a setting entirely organized around producing reliable witnesses.

It is thus a question of taking the risk of going beyond the notion of artifact, of learning to venture beyond territories mapped out by the imperative of proof and haunted by the denunciation of artifacts. And of doing this knowing very well that we live in a world dominated by this Dostoyevskian order-word: "If proof does not exist, everything is permitted."[9]

Let us begin prudently by a "real-life case." During a colloquium, in front of twenty or so participants, a cognitive psychologist proposes to show a film. He announces that we are going to see a dozen people, half in white T-shirts, the other half in black T-shirts, passing balls to each other. We, the participants, are asked to count the passes between the players wearing white. So we try to do that, and it isn't easy. But, when the film is over, the psychologist shows no interest in the numbers laboriously extracted. He wants to know if we noticed anything special, and confronted with a unanimous no, he shows us the film again, this time with the advice to "simply look." We find our players again, but suddenly we see a thirteenth person insinuating himself calmly among the players, who continue to pass the ball as if nothing was happening. He stops in front of the camera, waves hello, then goes off, just as he had come: none of us had seen him, and yet he was rather striking, disguised (in black!) as a gorilla.

What did our psychologist want to demonstrate? That conscious perception is a function of mechanisms that carry the label "pay attention."[10] Cognitive psychology thus valiantly pursues the enterprise that makes scientific truth and narcissistic wound converge: you believe that what you perceive is what is there to be perceived; however, we know, that this is not the case, that perception is a function of a non-conscious selective processing of information.

First remark: this has been known ever since there were illusionists and conjurers. The situation has a scientific allure because the accent is not placed on the successful, crafty diversion of attention but on a transparent but infinitely more precarious proposition: execute this task. I imagine that no one would dare to try the experiment with a class of boisterous adolescents, disinclined to do what they are told, for a single exclamation would suffice for all of them (and us) to see the gorilla.

Second remark: the lover of the artifact is in his element. We suspected that we were going to be duped—it's always like this with psychologists—but we also knew that it was part of the presentation, which we felt bound to follow as good participants in the colloquium: we thus felt obliged to play along. Playing along, in fact, wouldn't be any more bizarre than listening, while stuck in our seats, to a paper that we could just as easily have read. And it would perhaps have more surprises in store for us. But how far did our compliance in shouldering the proposed role go? To what point did we find ourselves obliged by the instructions? To go as far as abstracting out from what we nevertheless perceived very well, just as the hypnotized of the nineteenth century abstracted out from what proponents of role-playing today consider they remember very well? All the ambiguity of the testimonies about hypnosis are here, scarcely concealed by the fact that no one doubts what the exhortation "pay attention!" means, and that attention is therefore imposed like something that, incontestably, exists. There is no place for simulators: counting passes demands attention.

Third remark: whether or not it's a question of an artifact leaves me completely indifferent, and I don't feel the least bit duped, but, rather, content to have "consented." On the other hand, what does strike me is the contrast between the scientific presentation—here is a case exhibiting the processing of information of which this perception, which you believe to be a simple reflection of a given world, is a function—and the admiration that I felt for the great achievement, whatever it may be, that the adherence to instructions obliging [us to] a greatly concentrated attention made possible. Because we did not "perceive"; we had consented to confer on our speaker the power to transform us into detectors, interested only in white T-shirts, and, more precisely, in only a single question: To whom will the ball held at this moment by a white T-shirt go? We succeeded completely in eliminating everything else, which was only parasitic given the given task that absorbed us. The notion of "information processing" skips over what allowed it to be exhibited: consent.

The experiments that cognitive psychologists ask to exhibit a "function" are always presented as purified cases the only particularity of which is to permit an aspect of what is always the case in the "cognitive treatment" of data to be pinpointed. But these experiments imply either "traps" such as "hidden cameras" or situations that presuppose the efficacy of the kind of "paying attention" that makes subjects capable of producing the performance demanded. In both types of cases, they presuppose what "information processing" as such does not permit us to understand: the determined adherence of the subject who consents to the role proposed by the situation.

Of course, a supporter of the "new hypnosis" of Ericksonian inspiration, for whom all "mystery" has been dissipated, will snigger, because for him "paying attention" is part of the ABC of linguistic pragmatics, with its modes of enunciation: descriptive, self-referential, paradoxical, injunctive, explicitly or implicitly implicative, performative, and so on. But when linguistics classified these modes, it was interested with familiar cases, in a redundant connection with an instituted situation, with predefined roles: "can you pass me the salt?"

or "I now declare you man and wife." The instruction that made us able to "disregard" the gorilla, or the one that induces a talented subject to submit to dental surgery without flinching, exhibits, dramatizes, and vectorizes (constitutes as producing remarkable effects) distinctions that language in the normal sense makes function like Monsieur Jourdain produced prose. The linguist may calmly describe these distinctions, without being subjected to their efficacy (I speak here also as a philosopher, for philosophy is also an art of experimenting with language as power).

In other terms, if we put the accent on the performative character of language, we don't explain what it is we become capable of any better than we do with the cerebral black box of information processing. Linguistic analysis of induction ends up instead with language itself "changing nature," becoming what may connect us with a fabricating, creative power: maybe as fearsome as the gods and spirits "practiced" by those populations that we call "fetishistic." The spoken injunction thus does not "explain" the invisible gorilla without this invisibility obliging thinking differently about language, as inseparable from a power that is not susceptible to being subjected to any theory at all, of communication or anything else.

What would happen if, instead of seeking to "explain," to reduce experience, perception, consciousness to a function, we tried to encounter them on the basis of what binds them, on the basis of what is certainly a "cause," but not in the sense of a cause that permits explanation, but in the sense that a cause is associated with an efficacy, making us think, making us feel, making an experience exist in one mode and not another. We would at least have succeeded in making explicit the difference that I tried to make in the first part of this little book, between scientific success and the catch-all notions that give scientific progress the appearance of a uniform expansion. For if there is one practice that clearly exhibits the distinction between "explanation" and "cause," it is that of the scientist who is obliged to think by what he is dealing with, who is literally "called into question," and sometimes rendered capable of the type of event that is called a leap of the imagination (a description as good as any other when the term "imagination" is connected to the question of becoming, not to a "catch-all" psychological function). On the other hand, the catch-all notion is not a "cause," because the scientist armed with such a notion already knows what he must recognize, what he must judge.[11] And we would equally gain a new perspective on the history of hypnosis, relieved of the judgment that "it's all mere artifact," for what counts, from now on, is the "production of facts of art," of situations where, because of what is "done," the subjects, "simulators" included, effectively become capable of what they would not have been able to do without it. For better or worse: let's remember the Milgram case.

What have we accomplished via this example? We have passed from a "scientific" question (can such and such a phenomenon be exhibited as "a function of," as "permitting explanation by") to a question that designates first of all the successful production of a transformation, whether that is a matter of us, obliged by our adherence to the instruction, rendered capable of not seeing the gorilla, students of Milgram rendered capable of torturing their peers, or simulators made capable of playing the game so completely that they dupe their scientists. But in doing this, we have also changed practical domain. For the construction of knowledges that aim at such productions, that seek to stabilize them, that experiment with circumstances and components, belongs to "technicians" not to "scientists."

Unlike experimental practices, technical practices escape all opposition between fact and artifact, reliable witnesses and compliant witnesses: this opposition only makes sense if the vocation of the fact is to designate a respondent who authenticates its testimony. No one would think of accusing a technical apparatus of "fabricating" what it produces, or of creating new meanings. The "success" of a technique has no relationship to truth values (which does not at all signify that it is relative to a simple "it works").[12] Correlatively, while the notion of "science" in the modern sense cannot or should not be generalized to modes of knowledge cultivated by other traditions, there is no reason to restrict the notion of technique to our "modern" practices. This would only be so for "scientific techniques," referring their intelligibility to beings resulting from scientific practices. And the question that concerns us is thus one of freeing ourselves from the great history of progress according to which every technique is called upon to become "scientific."

Even with regard to a technique as obviously transformed by experimental chemistry as metallurgy, the tale that dramatizes "progress" as it goes from traditional techniques to a technique that is "finally scientific" is partially false. In effect, it induces the idea that one can assimilate the finalities of "scientific technique" and of "traditional technique." Yet neither "scientific technique" nor "traditional technique" respond to the single abstract finality of "producing useful alloys." The traditional blacksmith was no more a simple metalworker than modern metallurgy is a simple ensemble of processes for producing ever more sophisticated alloys. These processes abstract out from an immense network of practices, institutions, conventions, and new questions, such as overproduction. In other words, "purification," the elimination of "rituals," of "customs," of "beliefs" judged parasitic, is always a partial and biased description. All "scientific redefinition" is equally a change in the nature of the technique.

But we can go one step further and ask ourselves if "modern medicine," armed with clinical tests, has attained the envied status of a technique that is finally scientific. The necessity of clinical tests betrays the fact that, in general,[13] it is nothing of the sort. What is "proven" is that a molecule which succeeds is part of the cure, but the necessity for a statistical test to link the molecule obtained by the scientific research that precedes it and the molecule endowed with curative efficacy that follows it, shows the inability to make this efficacy testify to a finally scientific understanding of the cure. That does not prevent the reference to science from playing an important role in modern pharmaceutics. But this role has no autonomy; it is caught in a "machinic" functioning that correlates industry, the market, different governmental regulations, the eternal struggle between doctors and charlatans, and so on.[14]

Metallurgy is, rather incontestably, a technique which I will call, not "scientific"—because science is not in control—but "modern" nevertheless, in the sense that the symbiosis between scientific, technical, industrial, commercial dynamics doesn't need scientific authority to state its difference with traditional metallurgical techniques. On the other hand, one may talk about pharmaceutics as a "modernized" technique, organized in order to present itself as a technique that is, finally, scientific. And it is the same for Freudian psychoanalysis, the explicit mission of which was to attain the status of a modern technique, in which the efficacy of the cure would respond to a finally scientific reading of the psyche.

What does this notion of a "modernized technique" provide? It is not a question of a denunciation but of a proposition aiming to create distinctions that liberate thought where it is weighed down by the division between "before" and "after." The order-word that judges a technique according to whether or not it can exhibit its scientific justification thwarts any relevant characterization of techniques as such since the stamp "authorized by a science" makes the commanding difference and is opposed to thinking techniques in the full development of what they entail, the new problem they create, the allies or other concerned protagonists which will contribute to their meaning.

Let us take an example calling forth such a deployment, that of contemporary laboratories within which, first with primates, then with parrots and other promising species, researchers try to learn what an animal can become capable of. In *Quand le loup habitera avec l'agneau*,[15] Vinciane Despret has shown that the two statements "the animals have changed" and "their specialists have changed" are, in this case, inseparable. Because it is a matter of learning to address primates or parrots in a way that interests them, making them partners in an apprenticeship. Not so long ago, Harlow and many others systematically tortured their "subjects" in the name of scientific proof.[16] Harlow made innumerable rhesus monkeys go "mad" in order to "scientifically establish," with the support of statistics, that their pathological adult behavior was indeed a function of what they had endured when young. There aren't any statistics for those who, on the other hand, no longer work with a specimen representative of a given species, but with a particular being embarked with them on a "trans-species" adventure. Correlatively, nor are learning situations subject to the imperative of turning the animal into a reliable witness for a capacity attributable to its species (abstract tasks supposed to demonstrate, for example, that chimpanzees are capable of formal-logical reasoning). What prevails, what makes a situation interesting, is the animal's interest, the manner in which it can take the initiative, put different aspects of the situation into play on its own behalf.

From the point of view of the question, "what is an animal?" or "to what does the psychic functioning of an animal submit?," what is produced are obviously artifacts, performances of "denatured" beings that are not representative of those who gambol about "in nature." But if one situates this research in the filiation of the much older adventure of techniques of domestication, they become new and important. Domestication has succeeded in obtaining from some animals that they become inhabitants of a human *domus*, but this success was above all described as a function of the services that the animals became capable of rendering to humans. In our case, on the other hand, the question is deliberately open: what are they liable to become capable of in relation to situations they had never encountered? Situations where they are relieved of what, in their normal environment, demands that they be ceaselessly mobilized, on the look-out: they are protected, they have shelter and food. But they are not "free," as such, but rather held in an "artificial" environment, thought up in order to propose and make observable behaviors which interest "us," which oblige "us" to recognize in them capacities that "we" had denied them.

The questions "what can it become?" and "what can it become capable of?" therefore have nothing to do with the romantic ideal of becomings being spontaneously produced when a being is freed from what limits or alienates it. What Kanzi, Koko, Chanktek, and Alex—no longer "a" bonobo representative of his species, or "a" gorilla, "an" orangutan, "a" parrot—succeed in doing matters for passionate and dogged researchers coupled with their animal in a way that may evoke the magnetizer/clairvoyant coupling. This includes the fact that the performances elicited may be contested, disqualified, dismissed as misleading by every inquisitor as dogged as those with whom the clairvoyants had to deal (the "Clever Hans" effect).

But are the places where these new beings are created still laboratories? Perhaps not, but perhaps they are in the sense that what is learned there interests many others than those "colleagues" for whom "what is not proven has no value."[17] Researchers here are in the service, not of the proof but of the possible relation to stabilize and nourish, in the service of a cause important enough for the fact/artifact opposition to lose its value as a touchstone. That is why Vinciane Despret places her study under the sign of a prophecy: at the hour of peace, Isaiah announced, "the wolf will live with the lamb." If peace comes one day, these new beings will have helped us entertain new ways of considering those that, today, we exploit, torture, massacre.[18]

The fact that the researchers that I have just talked about are recognized as "scientists," even though they are prolonging and renewing the ancient art of domestication, is fortunate for their research, since it is widely understood today that making science constitutes the only legitimate way to learn. In this case, they are protected by the stability of the biological definition that "we" confer on animals. They can enter into a process of singular and creative co-construction, but they can still present themselves with the label "we used to believe that chimpanzees or parrots . . . now we know that . . .." And it is exactly this stable reference point, "it's still a chimpanzee" that matters (although it doesn't translate into a "reliable" response to the question "what is a chimpanzee?") and which is missing in the history that concerns us. If a reliable marker, which would guarantee that a subject is "truly hypnotized," could have been defined, perhaps researchers in hypnosis would have been able to gain the freedom of their ethological colleagues: the freedom to concoct situations that intensify, which make possible, the freedom to "explore with."

It is important to stress that to speak about a reliable marker does not in the least coincide with the kind of success that makes the opposition of "fact" and "artifact" possible. The marker must not be confused with an "objective" definition (the drawing of an encephalogram does not tell us what the dream "is" in the sense of "what questions it answers"; it only guarantees that a subject is indeed dreaming).[19] It is a question of a specific characterization, which cannot be simulated, and that I would here call "conventional," but in the strong and positive sense of the term. The possibility of such a characterization does not imply the submission of what is characterized to categories of experimentation, but, on the other hand, it does permit researchers to *convenire*, to come together, around subjects without being annoyed or divided by doubt. It would make possible the tranquility with which we speak of Koko as a young gorilla, free us from the insistent question: Are we dealing with subjects who are "really hypnotized" or with subjects compliantly playing the game named hypnosis?

If modern pharmaceutics can present itself as authorized by the success of clinical trials, it is owing to such markers, which allow an illness to be identified and an improvement or a cure to be evaluated. The trial that the clinical test stages depends on such markers as much for the constitution of the statistical sampling of sick persons as for the evaluation of the results, rates of cure or improvement.[20] The clinical trial has the power to "bring together" researchers, business people, administrators, journalists, stock-market speculators, doctors, who all await the result, because they all accept the legitimacy of the verdict. The marker assures the stability of a "convention" without which each and every one of them would be vulnerable to the accusation that they themselves bring to bear on "charlatans": that of participating in an enterprise profiting from the credulity of the public and from the demand for help coming from those who suffer.

The situation does offer some exceptions. One thinks on the one hand of antibiotics, the effectiveness of which answers to the experimental ideal (it is easier to define what "kill the invaders" means than it is to define "heal"), but on the other of the annoying case of psychotropic drugs. In their case, the "troubles" are characterized solely on the basis of the testimony of sufferers, and one has, moreover, every reason to think that this particular characterization is a co-construction particular to our time, a "cultural artifact," one might say. The efficacy of molecules is thus not anchored by any other convention than that of consensus, and what makes the ensemble "hold together" may well be the dynamic of innovation, first of all, every molecule always being the "the next to last."[21]

The example of psychotropic drugs says a lot about the thesis I am presenting here with regard to the drawbacks of the notion of a "modernized technique." Whenever it is presented in this way, a technique depends in a crucial manner on the existence of a reliable marker, which permits it to affirm the validity of its definitions, and thus its close relationship to scientific practices. Being presented as products of a technique that is "finally scientific," psychotropic drugs invite us to think, in a mendacious way, that "mental troubles" have finally become an object of science. If, on the other hand, one considers them as resulting from a technique that does not, in this case, benefit from the definition of a reliable marker (any more than does hypnosis), but that may indeed induce some relief from what patients complain about, there is no reason to denounce the situation but rather to describe it as "beyond science," in the pragmatic mode that any therapeutic technique calls for.

However, most of the protagonists involved in the bringing of psychotropic molecules to the market act "as if" they were using a reliable marker, even a scientific explanation of the efficacy of their molecules. They can do so because no one has either the interest or the power to break this consensual fiction. Those who try it come up against very considerable indifference: they are barking while the caravan of new molecules passes by. On the other hand, in a history like that of hypnosis, the absence of a reliable marker weighs heavily. If it had been possible to characterize the "hypnotic state" in the way states of wakefulness, deep sleep, and dreaming are "conventionally" characterized, the controversies would doubtless have been able to "bring together" instead of dispersing. Consenting to hypnosis might have been considered as forming part of the setting, a condition making possible the production of a state or of a psychic regime having the power to guarantee that researchers with different procedures are studying the "same" phenomenon. They would then have been free to vary the modes of induction, the situations, the manners of addressing the patient, of multiplying the facts-artifacts, because their technique would have been protected from the suspicion of fabricating dupes, of encouraging simulators, of making up a reality "that doesn't exist."

The same curse, the absence of a reliable marker, also weighs, let us note, on "parapsychological" research, which directly inherited and sought in vain to modernize the extra-lucid faculties cultivated by the magnetizers of the nineteenthcentury (hypermnesia, clairvoyance, extra-sensorial capacity to see and hear). Independently of the prevailing skepticism, the researchers themselves are besieged by the suspicion that they could be victims of fraud.[22] And the conditions of control that they impose upon themselves to fend off this suspicion become ingredients of what they try to stage, for these conditions comprise a very clear message destined to the "gifted subject" who lends himself to the experiment: he must accept an environment that defines him as suspected of fraud. A definition of the phenomenon by suspicion has replaced the relationship of trust between the magnetizer and his clairvoyant, confronting a skeptical world together.

Obviously, accepting hypothetically the contingent character of the absence of a marker for hypnosis can be contested and will be so by all those who hold that hypnosis "is only" role-playing. I accept the hypothesis of this contingent character not because I "believe" in hypnosis or, indeed, magnetic clairvoyance but because the difference is crucial only in the case against which I am trying to think. Today, techniques have no value, do not matter, and cannot bring together, unless they can present themselves as "modern," deriving from a finally objective characterization. But for anyone interested in becomings, whether of chimpanzees or parrots or the clairvoyant becoming of mediums, the question is secondary, what matters being what a technique brings to existence, and not what it testifies to.[23]

From the point of view that I am defending, the criticism of hypnotism by magnetizers is legitimate. The hypnotic redefinition of magnetism translates the ambition to purify that designates practices haunted by their necessary "modernization." For anyone interested in "causes," in the obligations these causes entail and the becomings associated with them, it can no longer be a question of purification but of the transformation of the "cause." And this transformation includes the difference between what the magnetizer and the hypnotist anticipate. The first addresses his clairvoyant man or woman as one who is able to become a vehicle, a medium, for forces that exceed what we usually call nature, while the second addresses the hypnotized person in so far as her artificially induced behavior will constitute a possible variation of a psychic function common to all humans. The difference is remarkable. To question the behaviors of hypnotized subjects in the laboratory, it is sufficient to recruit a bunch of students and ask them to play the role of simulators. But to contest the extra-lucid talents of the magnetized, it was necessary to attribute to them reputedly "normal" but nevertheless astonishing talents, for example an extravagant sensitivity to "signals" imperceptible to everyone.[24] And in order to try to "undo" the "paranormal" facts associated with their descendants in parapsychological laboratories, it is conjurors, those expert in all sorts of fakery, who are recruited.

Let us add that, to undo the authority of the multiple personalities' memories, delivered as they were with impressive sincerity, their therapist-fabricators had to back away in the face of the threat of lawsuits and abandon their now false witnesses and their henceforth "false" memories. But this is not really an objection, quite the contrary, because first of all it singles out the formidable cause constituted by the "proof" in the name of which these therapists and their subjects entered on their crusade. To the extent that it authorized a short circuit between all categories—therapeutic, scientific, and judiciary—proof has swept away all the safeguards thanks to which other cultures can have recourse to a somewhat similar practice, that of explanation by witchcraft/sorcery. Correlatively, denouncing proof simply as an artifact allows the page to be turned with no further thought, leaving "victims" only the possibility of "changing sorcerer," of pursuing in court therapists henceforth held responsible for their bewitchment. The categories of proof and artifact create obstacles to the thinking to which technique appeals: certainly "it works" but what matters is "how."

To a certain degree, hypnosis works, whether one refers to the ensemble of techniques experimented with a little everywhere, from top-level athletes to senior executives in multinational corporations in order to improve their performances, to "liberating creativity," to the current use of relaxation techniques to obtain more intense possibilities of concentration, and so on. But the history in which I am trying to intervene was marked in the first place by the project of conferring a common identity on manifestations that are, in Europe and for many peoples around the world, associated with "causes" that are irreducible to humans. It is not right that this project, which took part in the grand history of scientific progress, should end up in a mosaic of techniques, freed from their dream of modernity, and unified by a simple "it works." For the moral of the history would then, for its part, remain essentially intact: we would have nothing to learn from "non-secular" practices, they would not oblige us to think. They would only exemplify the heterogeneous gamut of beliefs and superstitions associated with techniques that, for their part, "blindly work." The slippery slope that has to be climbed back up, after the accusation "these are only artifacts" is abandoned, is precisely the one that has led us to this astonishing judgment: apart from science, there is nothing to learn, nothing to think about. It is enough to say, "it works."

# Notes

1 In both cases, the most intense requirements weigh on the quality of statistical analyzes, on the possible "biases" introduced by the questions, and so on. But what to do if the interrogated subjects who obligingly lend themselves to the game of dupes or to the role that the questionnaire assigns them, know what is expected and play the game; that is to say, they refuse to deceive, to clash, to irritate? Léon Chertok knew well the experiments in which hypnotized subjects, and also simulators, were asked

to produce gestures supposed to endanger the physical integrity of participants (see on this subject, Delbœuf, "M. Liégeois et les suggestions criminelles," as well as J. R. Laurence and C. Perry, *Hypnosis, Well and Memory* (New York: The Guildford Press, 1988): all perform these "criminal suggestions," and those who were interrogated afterwards affirmed that they knew well that they were in the laboratory and that, in one way or another, their gesture would not have the announced consequences. For Chertok, the famous experiment of Stanley Milgram, where subjects mutated into torturers sending ever more intense electric shocks into screaming participants, could not sidestep this doubt, even if the attitude of the subjects seemed to guarantee that they indeed thought they were conducting torture.

2   It is doubtless because he was familiar with the hypnotic tradition that the analyzes Bergson consecrated to the explanatory pretensions of psychology have lost nothing of their topicality. One can say that his *Time and Freewill* (1889) submits the way in which we normally describe the "data of consciousness," including the way in which we try to explain ourselves, to analyze our hesitations and our choices, to determine our reasons and to explain our decisions, to a reframing that is nothing other than that of the hypnotic phenomenology that Delbœuf initiated.

3   See Ian Hacking, *Rewriting the Soul. Multiple Personality and the Sciences of Memory* (Princeton, NJ: Princeton University Press, 1995).

4   Daniel Dennett, *Consciousness Explained* (London: Penguin Books, 1993), p. 419.

5   See, for the case that began the epidemic, Mikkel Borch-Jacobsen, "A Blackbox Named Sybil," in *Making Minds and Madness*. One will note that the production of a genuine multiple personality, clearly displayed (with often impressive talent), takes time, like the creation of the magnetizer/subject couple takes time.

6   See Susan Blackmore, *The Meme Machine* (Oxford: Oxford University Press, 1999).

7   See Vinciane Despret, *Our Emotional Makeup: Ethnopsychology and Selfhood*, trans. Marjolijn De Jager (New York: Other Press, 2004).

8   See Philippe Pignarre, *Qu'est-ce qu'un médicament?* (Paris: La Découverte, 1997).

9   This domination found new fields of application with the henceforth open question of the evaluation of so-called alternative medicines. The ensemble of what will follow, and notably the necessity to learn to think a technique as such, and not on the basis of its reference to a science supposed to justify it, is, in this case, profoundly topical. And on the other hand, there is the remarkable constancy with which, having condemned these techniques because they do not satisfy proofs that define contemporary medications, authors speak of the interest, even urgency, of a "finally scientific" study of the therapeutic efficacy that alternative medicines manifest cannot be vindicated because their claims are not scientifically founded. Such appeals to urgency have resonated in literature for more than two centuries.

10  I exclude here the hypothesis that the psychologist might in fact have demonstrated our membership in the scientific community where one is not supposed to cheat,

since none among us doubted the fact that he showed us the same film again. If this weren't the case, we were dealing with a scientist using a conjurer's trick to accumulate data on the credulity of researchers.

11  This distinction overlaps the two interpretations of Thomas Kuhn's notion of a paradigm. According to several interpreters of Kuhn, the paradigm functions as a catch-all: scientists extend from case to case a particular "way of seeing." But according to Kuhn himself, the paradigm extends through the successful solution of what he calls "puzzles," or brain-teasers—not simple judgments of resemblance but demanding exercises because each case must effectively reproduce, although in a minor mode, the type of success to which the paradigm owes its existence, the production of a new reliable witness.

12  In the next chapter I will approach the delicate question of an evaluation of therapeutic techniques. In a general manner, let us say that it is posed in terms of the interrogation enunciated by Bruno Latour in his *Politics of Nature*, trans. Catherine Porter (Cambridge, MA: Harvard University Press, 2004): Do we wish to, or do we accept, to live with beings produced by this technique? Let us stress that this question is always concrete: a GMO created by Monsanto, for example, is not the same being as the possible GMOs that might one day be created by research led in a concerted way with concerned Indian peasants.

13  I exclude notably the case of antibiotics that succeed in killing living organisms without doing much harm to the body that these organisms invaded: this triumph of medicine has the particularity that the healing is here reduced to the elimination of what is responsible for the disease.

14  See Pignarre, *Qu'est-ce qu'un médicament?*.

15  Vinciane Despret, *Quand le loup habitera avec l'agneau* (Paris: Les Empêcheurs de penser en rond, 2002).

16  I am part of those cohorts of students who, in the framework of a course in experimental psychology, "learned" Harlow's experiments like they learned about the territorial behavior of sticklebacks, without the least bit of scandal: facts . . .

17  The great difference between ethologists and modern therapists, from this point of view, is that every modern therapist whose treatment would result in the production of persons who claim "clairvoyance" would be viewed in a rather bad light, even if these patients are talented and happy with their "gift." This is not socially accepted. Today it is legitimate to create new beings with animals, liable to confer a meaning on what had none for their fellow creatures, but not to explore the links between healing and becoming. Healing is supposed to signify becoming normal (again) (or becoming a psychoanalyst, it will be added after Lacan).

18  Science-fiction author David Brin, a contemporary of this research, prolonged it in an unexpected way. A cycle of nine novels [in his "Uplift" series TN] imagines a future when humans would have dedicated their efforts to "uplift" other species by all sorts of manipulations. Two promising species, chimpanzees and dolphins, are raised to the

status of "*sapiens sapiens.*" In doing this humans believed they were "innovating" in relation to biological evolution, which they thought had permitted humans to "uplift by themselves." But, after the "first contact" with intergalactic civilizations, they were forced to realize that they were indeed the only ones in the universe to believe that a species could do that. For all the other inhabitants of this fraction of the universe, humans are a sort of "wolf children," abandoned while young and thus very badly behaved.

19  Without such a marker the possibility of "lucid dreams," in which the dreamer, after a long and demanding training, becomes capable of deliberately directing the course of her dream, would remain of the order of "unverifiable anecdote." The fact that the trace of encephalograms shows in a reliable manner that the subject is indeed dreaming, even though she is communicating with the experimenters according to a predetermined code, made the difference. The experimenter knows *by other means* than her own testimony that the subject is indeed dreaming, and the established facts thus avoid suspicion of fraud, conscious or not.

20  Steve Epstein, in *Impure Science. AIDS, Activism, and the Politics of Knowledge* (Berkeley and Los Angeles, CA: University of California Press, 1996), has superbly showcased the particular case of AIDS: while none of the protagonists of this slice of history doubts any longer that "viruses are the objective cause of AIDS," the question of markers permitting the evaluation of the efficacy of candidate therapeutic molecules made the patients intervene: they refused the one "truly reliable" marker proposed by experts, that is the death of the patient.

21  See Philippe Pignarre, *Puissance des psychotropes, pouvoir des patients* (Paris: PUF, 1999); *Comment la dépression est devenue une épidémie* (Paris: La Découverte, 2001); and "Qu'est-ce qu'un psychotrope? Psychothérapeutes et prescripteurs face aux mystères de la dépression," *Ethnopsy* 2 (March 2001): pp. 241–92.

22  Here again, it is science fiction that is the most lucid. In *The Ride of Pegasus*, Anne McCaffrey imagines the consequences of a particular encephalographic trace coming to be associated in a reliable manner with a "psy event." This is a technical event, not a scientific one, because the trace permits a reliable correlation, not explanation. But it will liberate "gifted subjects" from the doubts that were poisoning them, permitting them to reassemble and cultivate their gifts together. And finally . . .

23  This was the position of André Breton (see Méheust, *Le Choc des sciences psychiques*, volume 2 of *Somnambules et Médiumnité*, pp. 317–40) who proposed the freeing of procedures associated with the production of lucid consciousness associated with magnetism from the hands of both scientists and physicians in order to cultivate what is important in the creation of poets.

24  As Méheust remarks (*Le Choc des sciences psychiques*, volume 2 of *Somnambules et Médiumnité*, pp. 163–74), the hypothesis of an exaltation of sensorial and intellectual faculties of "somnambulists" is purely tactical: it serves to disqualify claims and is "forgotten" as soon as the operation has been carried through to a successful conclusion, even though one might naively have thought it worthy of interest in itself.

# 6

# *Thinking Therapeutic Techniques?*

Today, the project of conferring a common identity on the ensemble of manifestations that magnetism and then hypnosis had the ambition of bringing together has not been abandoned, but it does have a seemingly more neutral formulation. It is a matter of "trance" phenomena. Hypnosis, like magnetism, like possessions, convulsions, and maybe even like any relation of adherence to a role, would be so many different actualizations, referring to different contexts, of this common trait. "Trance" would thus constitute a common reference able to bring together not only the trances that ethnologists recognize as such, ritual trances of possession, for example, but also "spontaneous trances," mystical ecstasies, or the techniques of Zen meditation, and finally perhaps the ensemble of those behaviors that animal ethologists study, which were previously called animal hypnosis.

The notion of trance, which different techniques give the opportunity of manifesting in different modes, is most often associated with a key term: *potentiality*. It is with a potentiality, moreover, that Léon Chertok had proposed to associate the term "hypnosis"—which in this case he gave the generic range generally conferred on the term "trance." Potentiality, Chertok wrote, was "a fourth state of the organism, currently non-objectifiable (in contrast to the three others: wakefulness, sleep, dream): a sort of natural potentiality, having its roots in animal hypnosis, characterized by traits which apparently go back to pre-linguistic relations of the child and occur in situations when the individual is disturbed in his relation with the environment."[1]

It is important to stress that Léon Chertok did not surrender to the temptation to give the potentiality that he designated the least power to effectively bring together the multiplicity or to explain it. Following Lawrence Kubie, it was a question for him of affirming the refusal of a psychoanalytic, communicational or interpersonal definition of hypnosis, and of embedding the problem in the generic question of the relation with the environment: this "potentiality" would be a "common trait of all living beings but one to which each species would give a different content, proper to the type of relation that it maintains with its environment."[2] As to the fact of speaking of a "state," it was a question for him of refusing the reduction of hypnosis to suggestion or simulation.

Nevertheless, even for Léon Chertok, the notion of "potentiality" had a positive dimension, as if, somehow, it "said something" about hypnosis and other forms of trance. And effectively, throughout the history of magnetism, then of hypnosis, the same conviction is repeated, from interpreter to interpreter. The hypnotic rapport "would free a potentiality." But can one "hear" this conviction without giving to potentiality the possibility of an access to a "universal human potential" beyond diversity?[3]

For me, this question is decisive. Because with universal human potential, the essence of the modern project is preserved under the guise "we moderns know better": "they" believe that beings are showing themselves, possessing them, making them speak, dance, writhe in convulsions; we know that it only concerns particular manifestations of a human potential. The "only" is the mark of modernity: unlike others, we know that we are "alone in the world."

The formula "they believe . . . we know" has more than one trick up its sleeve. For it to impose itself, it isn't necessary that what "we know" respond to a finally scientific definition. What is essential, on the other hand, is the possibility of judging the past from a position that is finally purified, detached from the illusions that permit its definition. That is why it will be no surprise that this potential can find itself linked, not to a scientific procedure but to an initiatory procedure. It's enough for this initiation to take its model from Eastern meditation, which works to pass through the veil of illusions that make us adhere to the world.[4]

A very high-quality example of this position is proposed by François Roustang, who was initiated into hypnosis in the Ericksonian framework but did not stop with the communicational interpretation. Like Léon Chertok, moreover, François Roustang dreads the utilitarian blindness of "it works." Hypnosis must oblige us to think, and in this case, to think what it gives us access to. Consequently, beyond a diversity vulnerable to the statement "everything is valid," François Roustang distinguishes different "phases" of hypnosis, the access to potentiality as such "in its simplicity and its purity,"[5] constituting its final outcome and truth:

> induction begins by halting ordinary perception, for example in the form of an object independently of its context. The first phase, usually the only one accepted by opinion, authorizes seeing in hypnosis merely a phenomenon of fascination. But this first phase is only a passage. It leads to a second phase, of the suspension of the determinations that we are habituated to concerning things and beings. An indetermination that produces a feeling of confusion. The third phase is that wherein new possibilities arise because we have been freed from ties with the constituents of our existences that are too tight and self-evident. This is the phase when, for hypnotizers in the laboratory, hallucination may be produced. What remains is to reach the goal, that is to say, to attain potentiality as such, beyond confusion and imagination.[6]

From the point of view of access to potentiality "as such," the technical dimension of hypnosis, the "disturbance of ordinary relations to the environment," is secondary, reduced to a simple condition of access to a "beyond" independent of this technique and in particular independent of the confusion and imagination that it involves. Of course, for Roustang, this "beyond" has nothing to do with an object of science, experience of it is not to be "explained" or "operationalized in a function." It is rather a question of going beyond any explanation, any function, including the one that situates our experience in well-determined spatial-temporal referents. However, contrary to other initiatory techniques, the procedure that he describes is indeed modern in the sense that it communicates with a power to judge, to define the manner in which we must understand what we are living when we live in a mode claiming public validity. Pragmatically, it has in fact inherited from hypnosis its claim to organize under a single category the products of a parasitic imagination, hallucinations, beings that, for other peoples, what we call trance is thought to connect to. Roustang's potentiality may well reconnect humans to the cosmos; it conserves its very modern capacity to situate the beliefs of others.

It goes without saying that François Roustang's "potentiality" with its "purity" and its "simplicity" in no way escapes a reading that would make of it a *particular fabrication* of experience: a fabrication that would doubtless have as one of its ingredients the devaluing of the possibilities attributable to the imagination, since the "aim" is to access an experience beyond the imagination. And the ensemble of oppositions that populate Roustang's text, between pure, "naked," "animal" spontaneity on the one hand, restrained, calculating, self-interested intellectual mobilization on the other, would belong to the fabricating of this particularity as much as the ideal of acceding to spontaneous primordial purity does. In other words, in Roustang's text we again find the description of a trance, the establishing of which designates as its "cause" a "being," "pure spontaneity," endowed with the power to save us from the intellectual imprisonment that mutilates our lives.

There is thus strictly no reason to dispute François Roustang's technical and practical choices, nor the trajectory of those for whom it is effectively important to escape the urgency of mobilizations and commands. And nor is it a question of affirming that, if the experience of "pure potentiality" is not "primordial," it becomes a fabrication "like any other": that would be straightforward relativism. It is a question for me, on the other hand, of not surrendering to the satisfied "so, that's that," of not adhering to a proposition which would give its closure and its moral to a history that interests me by virtue of its stuttering. Resisting the seduction of this proposition, I will thus affirm that *like other techniques*, it poses the question of what it integrates: of what it supposes, of what it convokes, of the becomings set in motion by what it convokes, of what is said about it, concluded from it, testified to by it, and interpreted from it.

It thus is a question of learning to take seriously the efficacy of the techniques that put trances into play, that is to say, to accept the test that this efficacy inflicts on our "we now know" instead of reducing these techniques to the catch-all notions that permit these "we now know" to survive come hell or high water. Thus, one could certainly say that pure potentiality is only an "idea," but the question is then one of the power associated with the idea, which exceeds psychology (these are "our" ideas) like the power associated with a paradoxical manipulation of language exceeds pragmatic linguistics, as we have seen.

However, from the point of view of ethnopsychiatry, as proposed by Tobie Nathan, even the fact of speaking of "trance techniques" is somewhat dangerous, for the term "trance" invites us to think about what happens to humans "in a trance" and not about what the technician is addressing. To pose the question of trances, or of the trance, is thus to pose as judges, claiming to know the practice of technicians better than they do themselves, without even having to encounter them. We would not have to stop at the fact that, for them, what we call trance is first of all what connects with invisible beings and allows them to be negotiated with. We would have the right to separate their technique from their thought; that is to say, to think in their place. "Without knowing it," they would have cultivated a universal human potentiality, but it is up to us to separate it from beliefs we no longer adhere to.

Ethnopsychiatry, on the other hand, tries to take seriously the techniques of the practitioners of healing and does so by guarding against going "beyond," toward what would explain them "for us all." The art of the technician, which may be called one of arranging, selecting, convoking, stabilizing, in short of cultivating, does not put reasons into play but what I have called "causes." To these causes correspond a theoretical thought and the manipulations of objects that demonstrate its truth.[7] Of course, the articulation between theory and demonstration here does not respond to the demands of experimental proof, its aim is not to differentiate between fact and fiction but to induce a transformation. It is not a matter of silencing the skeptic but of obliging thinking in a mode that sets off the process of transformation. But, in order to be appropriately described, this process requires that what is set off not be reduced to a catch-all psychological category (identification, hallucination, suggestion, etc.). Taking seriously the efficacy of a technique imposes an understanding of it as addressing something more powerful than the technician.

This ethnopsychiatric proposition is a testing one. The stuttering of the history of hypnosis might suddenly be followed by the unanimity of a blunt rejection: if thinking the efficacy of therapeutic techniques cultivated elsewhere means having to take seriously the power of those beings that we now know do not exist, that we know belong to the domain of ideas, of effects of language, "we'll follow you no further."

Let us slow down a moment, because it is a matter here of thinking, not of rushing into a confrontation. It is a matter of learning how to abandon the space where the affirmation "this really exists" makes Koko the young gorilla and George W. Bush converge in the same mode of existence, permitting what is "only" an idea, an interpretation, an effect of language to be excluded (even if a war may then be waged between scientists and demystifiers of science with regard to the status to be accorded to Koko's and Bush's neurotransmitters).

What is language? What is an idea? Behind such familiar terms, it is the whole question of our syntactic usages that is in play, and no doubt that is the question that must stutter. The human being, in general, speaks and entertains ideas, we say with assurance, we who seek a bare definition, one capable of being valid beyond any cultural particularities of the human. But we also know that language is not a property of humans, a simple practical vehicle for transmitting information: words make us think and feel. In *Modes of Thought*, the philosopher Whitehead pointed out that the narrative of the sixth day, the day when God created Man, should be rewritten. It should read: "He gave them language and they became souls."[8] In Whitehead's sense, they became those beings for whom "what is" finds itself plunged into the possible, what could be, what will perhaps be, what one can imagine. On the sixth day, a creature of the possible sprang up, one who cannot be described unless this description implies that he/she is, or may be, prey to something more powerful than himself/herself. Our interrogations and our uncertainties are not "psychological," they are not ours, they are imposed on us by what could be, or what should be able to be.

Whitehead thus denies that, on the sixth day, man was created in "God's image," but likewise that the human is just a (rather) particular kind of primate. Those to whom language "was given" became souls, not "humans": one no more becomes human than Koko has to become gorilla. "Human" and "gorilla" are simply definitions, mute conventions apt to bring about the agreement of specialists. Except that for those specialists, to name ourselves "human" should make it possible to define "what is" beyond the diversity of customs, cultures, languages and divinities. In other words, Whitehead's characterization thus finds an exemplification but also the need to be completed. For it does not respond to the passion with which "we," among the members of the same species, claim to be "humans, only humans." "We," in this sense, are a bit special. In fact, Whitehead found in Plato what I think is the only non-"Eurocentric" definition of this category born in Europe, "the human." Humans, following Plato, are "sensitive to the Idea," and the power of the Idea is of an erotic type[9]: one does not submit to an idea, one does not obey it, one is "seduced," attracted, transformed by it. Even those who intend to reduce an idea to a particular configuration of neurons still bear witness to this power of the Idea; without it, why would they try to convince their contemporaries, to transform them?

It is not surprising that it fell to Whitehead, a mathematician before becoming a speculative philosopher, to fabricate utterances that require not that our definitions be reassuring but that they be complete, that their implications and consequences be followed through right through to the end. No, the Greeks—who had so much discussion of what the good definition of "man" is—were not simply the first to have adopted "the" normal position, the one that everyone should in turn come to adopt. They must be described as passionate, prey to a requirement stronger than themselves. Those who called themselves "men" have been captured by what they have baptized "idea" and this is what has made them "men."

And the history of the "science of man" called psychology consequently becomes rather curious. Scientists who have tried from the outset to eliminate the question of hypnosis by resorting to notions like "imagination" or "suggestion," have selected dismissive words relating it to a "merely psychological" explanation, but in other contexts, such words are precisely those which speak of the power of the Idea. And those who, in psychological laboratories, have tried to stage a hypnosis that testifies in a reliable manner to a "psychic functioning," and have clashed with the tremendous compliance of their subjects, have encountered the power that the Idea of collaborating with science, of adhering to the role that was allocated to them in the service of progress, has over these subjects.

Such staging is somewhat diabolic in that all the protestations it is fated to provoke will still testify to the power of the Idea that has made "men" of those who protest against it, a power that is fearsome, as are all powers, and, if not recognized, welcomed, cultivated as such, a power that may enslave and devour the soul of those that it obliges to think and feel. On the other hand, this staging agrees with the experience of those who know that what they seem to be the authors of is in fact what makes them exist. Do not mathematicians fabricate "objects" that make them exist as mathematicians and that are, moreover, liable to provoke what can be called trance states? A mathematician, of course, is not interested in trance any more than the shaman is: like the shaman, he is interested in the being whose presence the trance signals. And don't philosophers, certain philosophers at least, fabricate concepts that oblige them to undertake strange experiences, voyages out of the ordinary, that make them philosophers? And don't poets honor language completely differently than as something that obliges them without their knowing it? For all of these practitioners, the trance as such is not interesting, for they know well that it is connection, a mode of convocation of what makes them practitioners.

Other speculative philosophers have created other words. Hence for Deleuze and Guattari, a key term is "force," and every "assemblage" must be understood starting from forces that it captures, makes sensible or perceptible. The plant may chronologically precede the physicist, in the sense that for both solar light matters. But the songbird, caught up in the joyful song with which it "greets" the sunrise, makes this sun exist in a mode that is irreducible to a variable source of heat or electromagnetic radiation. It celebrates the sun in a mode that precedes the hymns of Akhenaten and the song of poets. Convoking, capturing, bringing into existence, being convoked, being captured, becoming, "none of these terms comprise the least danger of anthropomorphism or imply the least interpretation."[10] They make us think and feel against the anthropomorphism of functional explanations and against interpretations that permit us to oppose knowledge to beliefs.

Here also we again come across the manner in which Bruno Latour has characterized our curious passion for anti-fetishism. The fear of anthropomorphism, the threat of being convicted of interpreting instead of sticking to the severity of bare facts, signals us as "anti-fetishism addicts," addicted to the criticism of those we accuse of attributing an autonomous existence to what they in fact fabricate. And yet, the success of modern sciences signals the production of "fetishes" of a new genre, of factishes (atoms, neurons, bacteria). Experimenters attribute an autonomous existence to well-fabricated beings, beings that have overcome the tests their existence depends on for the experimental scientists: they have been capable of resisting the accusation of being a simple interpretative fiction and serve as reliable references for scientists.[11] Experimental proof may be called a "successful convocation," success corresponding in this case to the satisfaction of the requirements immanent to the adventure of modern sciences. Requirements that are equally "causes," which oblige scientists to think and dramatize the double force proper to all convocation: "The force of the one who knows how to convoke and the force of what might not respond to the convocation if it is done badly or if it gets its address wrong."[12]

So, what about the force of what I have called catch-all notions—symbol, motivation, suggestion, unconscious, communication, transference, cognitive treatment, and so on.—all of which affirm the ambition of bringing into existence psychic functioning as a scientific reference? It is important to recognize first of all that they have not achieved the enviable status of "factish." They inhabit "modernized" knowledges and techniques, not the high places of proof. But they are not for all that devoid of efficacy. The only problem is that they do not do what they claim and do not claim what they do. They claim to describe us, simply to describe us although their efficacy arises from the adherence that they elicit, and that designates us as veritable addicts of narcissistic wounds, to speak like Freud. "We are nothing but . . . " and the list is long: vehicles of our genes, prisoners of our language, products of our social habitus, results of our neuronal processing of information. "Yes, yes, that rings true, that looks true, therefore it is true . . . " And the adherence that they produce is expressed first of all by the proliferation of ignorant little judges, vectors of always poorer versions of the grounds that justify the condemnation of those who "still believe" without even having been listened to. If, with Deleuze and Spinoza, one names poison what separates somebody from their power of thinking, acting and feeling, the efficacy of such notions should be called poison.[13]

The singularity of Tobie Nathan's ethnopsychiatry, the reason why it is apt to oblige a non-clinician like me to think, is thus that it has conferred a site on such risky thoughts, coming from the speculative margins of our assurances, the site where they put something other than these assurances at risk. That is to say also the site where this marginality acquires another sense than that of criticizing prevailing positions. I feel obliged to think in the presence of those we have defined as having to be liberated from their illusory attachments, disqualified as beliefs, superstitions, and so on. I have to accept being challenged by those traditions for which the unthinkable thing par excellence would be that we are "alone in the world,"[14] that we alone are endowed with "intentions" in an indifferent world?

In fact, modern psychology can help us render the question a little less dizzying here, on condition of understanding its utterances in a manner that is complete enough to escape their triumphalist accents. In its project to imitate experimental sciences, modern psychology has indeed succeeded in undermining the figure that philosophers had opposed to objective reality, to the indifferent matter of scientific enquiry: the "intentional subject," the subject defined by the attribute of intentionality, which serves as a protection against the grasp of objective sciences. Far from being an attribute, intention would be a practical manner of characterizing what arises in certain situations, or assemblages, that convoke it. "What were you trying to tell me with that little grimace?": faced with this question, I immediately produce myself as endowed with intentions. All that remains then is not to leave the machinery of "we used to believe" running, but to affirm the singularity of a tradition that produced the "intentional subject." And to do so, for example, by referring to Michel Foucault's analyses of the techniques of Christian confession and the systematic and suspicious uncovering of hidden intentions to which those who bend to it deliver themselves up.[15]

Obviously, when the figure of the intentional subject prevails, the techniques of those who convoke invisible beings in order to interrogate their intentions force us up against the wall: "To believe in this or not?" Because intention then communicates with the entirety of a familiar psychological apparatus and invites us to conceive these beings in our image, to require of them that they furnish the same type of proof of the "truth" of their intentions as we have learned to require of each other. On the other hand, if we accept that our own identification with our intentions is the result of a long schooling begun in early childhood ("what do you want?," "are you sure that's what you really want?"), the question of "having intentions" is no longer a question of belief: the different modes of existence are never "in themselves" but always relative to apparatuses that convoke them or to assemblages that provoke them.[16]

From this point of view, it's not surprising that each time the truly experimental sciences, like physics, chemistry, or molecular biology, have progressed, they have identified new indifferent beings, which one can obviously not endow with intentionality. Every experimental apparatus requires the indifference of the witness to the stakes of his testimony, these latter concerning, and only concerning, the intentions of the scientist who poses the question. That was what the investigators in 1784 required of Mesmer's fluid, and it is what the clinical trials of the pharmaceutical industry seek to attribute to their molecules, but in a purely conventional mode.[17] As for the "catch-all notions," which typically permit the description of human behaviors in terms that allow their intentional character to be transcended, they mimic experimental success. In a general manner, each time that something presented as arising from a "subjective intention" finds itself reduced to the effect of an "objective reality," indifferent to our questions and our preoccupations, it is the same triumph against "opinion" that is celebrated. And we come to the great mysterious aporia of the *mind/body problem*, which asks: How can the necessarily indifferent neuronal brain "explain" our thinking, feeling, suffering experiences?

However, it is not a question of transforming the privilege accorded to indifference into a "cultural" singularity (exception). It can never be stressed enough that before serving to disqualify the "beliefs of others," the judgments that this privilege authorized were first effectuated in our own history, as an operator differentiating between those who know and those who are defined by illusion. However, it may be a question of evaluating the price paid by this "(anti-cultural) politics of truth," distributing what is legitimate and what is illusory. If so, we are in our element, for we are very talented at describing the "drama of the modern subject," or the sorrow of an "ever more disenchanted world." Passing from a description of something that "unfortunately we can no longer believe" to the description of something that "unfortunately we have irremediably destroyed" will not change the "white man's burden" a great deal. But ethnopsychiatry asks us a much more troubling question. The question is not one of what we can no longer believe in but rather concerns those questions we believe we can do without when we envisage techniques authorized by the objective "what is" (the neuronal brain, the unconscious, etc.). We may affirm the necessity of discussing the ethical desirability of some of these techniques, but we are not thinking of the technicians that they fabricate.

> Must I recall that, everywhere in the world, cultural systems have chosen to attribute to the therapist an ambiguous status, part savior, part sorcerer. I see two essential reasons for this. First, the real recognition of his activity, if he is capable of saving, he can also destroy [. . .]. Next, and especially, is the obligation to think therapeutic activity starting from the intentionality of the therapist. There is scarcely any society other than Western society that thinks its therapists on the basis of the soothing ideology of the good, describing them as knights of the health of humanity.[18]

To think intentionality, here, is not to speculate about intentions. The road to hell, Christians know, is paved with good intentions. Stanley Milgram did not have bad ones, things would have been so much simpler if that had been the case, and even Harlow, who tortured generations of young rhesus monkeys, was probably an honorable man. To think the intentionality of Milgram and of Harlow is first of all to think what they serve, what fabricated them, what commands them to do what they do. Otherwise, if one does without thinking this, one may oneself be "captured": sending electric shocks into shrieking accomplices or, as I did, learn "for the exam" how a torturer obtained what he called "facts." If psychology is a technique that presents itself as a science, it is necessary to remember that there is no technique without the fabrication of technicians, and to think the technique is to think what it is doing to the technician, what, more powerful than him, it puts him in the service of.

Perhaps the lack of robustness that characterizes the history of hypnosis designates the question of the fabrication of technicians. And perhaps the stubbornness with which many patients—in spite of their therapists—lend powers that are out of the ordinary to hypnosis would, in this case, express something that is completely different than irrationality: the knowledge that a technique that denies its own power and claims to explain itself in terms of "psychic functioning" is deficient. Gill and Brenman maintained, as we have seen, that hypnosis in fact poses more problems for the hypnotist than the subject, because it subjects the former to the temptation of attributing to himself a "magical" power. Or—when he thinks this power to be illusory, artificial—it imposes on him the test of having to maintain the comical role which made Léon Chertok say: "The hypnotist is like the Queen of England: she says 'I want', but it is Parliament, that is the subject, who decides." Or it obliges him to accept the worrisome turn that a cure may take when the roles are reversed, as was the case with Puységur (but not with Freud!).

It is from this point of view that the Freudian invention of the unconscious can claim all its interest: it has incontestably succeeded in stabilizing the respective positions of the analyst and his patient (also called the analysand). The analytic unconscious is not, as such, assimilable to a function (contrary to its "cognitivist translations" which reduce the conflict to a functional problem). It "performs" very effectively as a third party: it assigns roles, it inhabits the scene as both obliging and requiring. It is that the "message" of which must be deciphered, across short circuitings and slips of the tongue that distort what the subjects mean to say. The analyst is the one who knows how to understand its "asocial" insistence, irreducible to interpersonal significations and to the interests of life in common, that is to say, to the ensemble of variables that psychic functioning is supposed to articulate.

Like djinn or divinities, the invisible beings that enter into other therapeutic techniques, the Freudian unconscious has a power whose demand needs to be deciphered. Like those beings, it is what obliges the therapist to think as a therapist, that is, to inscribe the indeterminate suffering that he is dealing with into a space of signification that permits intervention. Like those beings, it confers on the therapist the role not of one on whom everything depends, but of someone who knows how to address what everything depends on. However, contrary to those beings, the unconscious is a being that was convoked by those who sought first of all to silence critics, to differentiate between "compliant testimony" and "proof" (and thus to confer on psychoanalysis the status of an "at last modern" technique). It is a contrast that has consequences that must not be underestimated. A being of a new kind.

In fact, the Freudian unconscious has inherited a rather fearsome quality from its strategic role in the question of the difference between proof and fiction. Convoked by an apparatus haunted by the imperative of making this difference, it is defined as indifferent, inaccessible to civilized commerce. And that is how it is honored. The indifference of the unconscious in relation to every negotiation is, for those who refer to it, the sign that we no longer believe, as others do, in a magical type of reality. It is the sign that psychoanalysis is indeed part of the grand history of rationality.

The Freudian unconscious affects humans as humans. It thus has the type of power that may be associated with the Idea, and more precisely—for ideas must of course be described on the basis of the becomings they give rise to—with those ideas that designate humans as having to become capable of stripping off the multi-colored rags with which they hide their nudity. But it differs from all other ideas in that it denies being an idea and presents itself as a being deprived of all Eros, that neither seduces nor attracts, that neither directs nor promises, a being "that is," purely and simply, devoid of intention, indifferent to the range, the signification, the consequences of its power. And it differs from the beings convoked by other therapists in that it escapes any possibility of negotiation.

From this point of view, it is no longer a question of opposing the lucidity of a Delbœuf to the coup de force of a Freud: they constitute two sides of the same coin, the coin that creates a value of generalized equivalence among all technical practices centered around "modes of commerce to be created," reducing them to so many fictions. And the warning of the magnetizers resonates again: hypnotism destines those who practice it to humiliate their patients, to see them as ill (Charcot) or suggestible (Bernheim), to transform them into living proofs of the practitioner's theory.

To have made it impossible to invoke the therapeutically interesting notion of "commerce to be established" or of "settlement to be agreed upon" (and to have reduced the "rapport" that magnetizers cultivated to a transference relation, disconnected from any becoming, liable only to be repeated blindly), is perhaps the drama of Freud, and of all those who tried to give therapy a scientifically respectable status. And doubtless, one of the reasons for the fascinating character of the Lacanian unconscious is that it has inspired the most radical and insulting slogan with regard to all commerce: do not give way on your desire. Desire itself gains the sovereign indifference that we oppose to the compromises of opinion. Correlatively, the link between analysis and "cure," already largely speculative for Freud, became an object of derision. Analysis fabricates analysts, beings who repeat after others what happened to them, discovering themselves to be "incurable," "irremediably split," learning to "know a bit about their desire," as they say . . .

It is always risky to adopt the perspective "of others," but when the fact that psychoanalytic practice has no "theory of the cure" is directly tied to the fact that the psychoanalytic unconscious is incapable of "commerce," a situation is created that is so curious that I cannot resist. For the point of view of these others, people who fearlessly frequent beings liable to invade a life, to devour a soul, and who convoke them without care, without practices defined as capable of keeping them at a "good distance," of creating protections or of negotiating peace, have, I understand, a name: they are sorcerers. And so the serious question then poses itself: Would not the incontestable efficacy of analytic technique, which actualizes and supposes the power of a "politics of the truth" that destroys fiction, have certain traits of the practices of sorcerers? Can the Freudian-Lacanian unconscious, indifferent and thus theoretically inaccessible to the civilized compromises that other "supernatural" beings (sometimes) accept, participate in anything besides practices that bewitch, capture souls thenceforth destined to work in its service? Analysis fabricates analysts, Lacan wrote. A fine definition for a sorcerer's machine because the sorcerer does indeed know how to create sorcerers.

Moreover, it is perhaps the analyst much more than her "analysand" who pays the price for her capture in an apparatus linking "truth" and the discovery of a blind and obstinate indifference, dismissing as fiction everything that analysis charges itself with undoing.[19] For the analyst is exposed by "modern" (but not scientific) psychotherapeutic techniques to ply her trade as "impossible," to borrow Freud's expression, and to dedicate herself to the service of this impossibility. As for the patients, they remain free to circulate.[20]

If one could risk using the term "impossible technique," it would certainly not be in the sense implied by Freud: a profession the impossibility of which is expressed by the fact that he attained the *nec plus ultra*, that he had rid himself of illusions, of false securities, of mythic interpretations. It would be in a much more perplexing sense. If our strange dedication to a truth that does not need to be cultivated (since it is marked by the power of conquering fiction) transforms age-old practices into "impossible professions," whether it's a matter of hypnosis or psychoanalysis, we do indeed confront a difference that all descriptions dismissing them as "therapeutic techniques among others" cannot smooth out: the beings convoked as having to manifest this power of truth are not among those with which a therapist can work.

The question of the evaluation of the efficacy of techniques that I have just approached is at one and the same time delicate—it is a question of avoiding the position of a judge armed with *a priori* norms—and robust, for it designates a matter of concern that doubtless can be found everywhere. I will call this question "pharmacological," for it addresses the distinction, which is unstable and thus calls for a stabilizing technique, between poison and remedy: "What is capable of saving can also destroy." The basic indeterminacy of the pharmakon, the problem of its handling, is part of our memory: Hermes was the god of merchants and thieves, and language, as we know, is the best and worst of things, and so on. One way of stating our difference in a mode that would not be tragic but would exhibit what "we believe we can do without" would thus be the refusal to pay attention to this question (except within the pharmaceutical industry). We are "those who do not believe in the power of our innumerable pharmaka."

As a philosopher, it behooves me to recognize the brutality with which philosophy has intervened here. Here where, in other cultures, it is a question of "paying attention," that is to say, of thinking, philosophers have sought the means for a disjunction that would in itself be stable. And it is this same stable disjunction that is promised by techniques that can present themselves as "scientific at last": the technique enacts "what is"; leaving "what should be" to be decided. But philosophy is not defined by this brutality. Everyone who has seen Gilles Deleuze's *Abécédaire* has heard him talk, in the hubbub of a lecture, of the possibility of every line of life being transformed into a line of death, a possibility that only prudence, attentive experimentation and careful protections can respond to. When he acknowledges the pharmacological art of becomings and their dangers, the philosopher says nothing different to what those healers who can be called "technicians of becoming" know, because, for them, what is called a diagnostic is first of all a divination: not of what is, but of becomings whose possibility haunts what is. "Diagnosing the becomings in each present that passes—that is what Nietzsche assigned to the philosopher as doctor, 'doctor of civilization' or an inventor of new modes of immanent existence."[21]

We live in a curious epoch. Many readers of the preceding pages must have concluded "all that is well and good, but we others, we disenchanted modern subjects, have lost contact with those becomings that we call 'healing.'" Yet, and this is the curious point, many readers, and perhaps the same, describe with great unease the proliferation and growing power of what we call "sects." And some will call any collective giving itself the means for "becoming-healing" that require a certain separation between "interior" and "exterior" sectarian. Let's not forget the stormy discussions provoked by those feminist groups who decided that some of their meetings would exclude men!

It is not a matter of "defending sects" but, as Tobie Nathan proposes, firstly and above all of avoiding easy explanations, which explain recruitment into sects in terms of the weakness, suggestibility, or disarray of those who let themselves be seduced. It is a matter of thinking sects on the basis of a force to which ex-followers, moreover, testify, and of which they are the living witnesses for as long as they have not managed to think it through.[22] And this thinking does not bear on the practices of humans, all too human in the eyes of the disappointed/deceived ex-followers. What must be realized is the "intentionality" proper to the sect. If there has to be a definition of sects, perhaps it should bear on the *ethology* of the beings for whom the sect recruits, what they require and to what they oblige.[23]

No more than is the case with chimpanzees, is the ethology of beings independent of the apparatuses that permit us to enter into commerce with them, the ways in which they are convoked as a "cause," the ways that are suitable for feeding and cultivating them in order to avoid being enslaved, reduced to the state of "zombies" by them. And ethology is not only central in questions of therapy. School, to cite just one example, is one of those places where "causes" are in play, which are supposed to enable students to think and to become, and if school today makes so many teachers and students "ill," it is perhaps because we are addicted to a "rational" definition of knowledge, as good and legitimate in itself, without granting the least attention to the art of convocations and the commerce to be established.[24]

The question of the "ethology" of beings is an eminently practical question, not a question of "belief." It is not necessary to "believe" in order to evaluate the power of mathematics to fabricate or destroy students, depending on the way it becomes "present" for them. Many adults are likely to burst into tears when they remember the humiliation of having had to become "automaths," to borrow Stella Baruk's apt expression, incapable of understanding what formulas "wanted from them," knowing that "they spoke" to others. As Bruno Latour has often written, our true difficulty is that "we believe in belief" and that when we are dealing with something whose power we have to learn how to characterize, we "psychologize" the problem. Hence it is not the power of mathematics that demands thinking but the "subject," his "lack of autonomy," the obstacles that he opposes to what it should be "normal" to learn. In the same way, when it is a question of therapy, one regrets the weak, easily influenced character of the "victims" of sects, but one also affirms that the individualistic character of subjects separates them from what it would be necessary for them to "believe" in order to make a technique work. Some believe too easily, others refuse to, but in neither case is the question of the force that an assemblage summons posed.

It is not certain that psychotherapists are in a privileged situation to take the risks that this situation imposes. They are a little too much "under surveillance," subject to sarcasm or the accusation of profiting from credulity like vulgar charlatans. They are in effect at the epicenter of what Foucault defined as the modern age of the history of the truth, at the point where the two "tectonic plates" that dispute its meaning meet: the "objective" truth that is conquered in the laboratory and the truth of the "subject" rendered at last capable of knowing himself. Therapists are incessantly called on to make common cause with one or the other. And patients are the hostages of this cause. Freud had no choice, if he wanted to avoid the accusation of getting his patients mixed up with his fictions, when he laid claim to the "objective" truth of the unconscious. Neither did the founders of hypnotism, if they wanted to avoid the scornful derision that engulfed magnetism. As for ethnopsychiatry, it survives in a hostile environment, subject to the accusation of confirming, even reinforcing, the beliefs of its migrant patients, instead of profiting from their suffering in order to free them from their traditional moorings, that is, from their "alienation."

That is why the unknown of the question that this little book attempts to characterize does not designate, primarily or exclusively, practices today defined as healing. If healers, up to and including the eighteenth century, were never very far from philosophers or sages, it is because their practices always captured what haunted an epoch. Not that these practices are a "function" of their epoch, but because they call for the resources and assemblages that give causes to the becomings an epoch is capable of fostering. It is thus the resources and assemblages that interest me first of all and in particular those which have the vital need of experimenting with new "causes."

Every reader will be able to think of certain practices, minority by definition. I have chosen to end this book by stating the importance, for me, of the groups that construct the concrete, pragmatic and, of course, technical means of escaping traditional modes of political mobilization. What is experimented with at the heart of these groups are apparatuses that allow them to cultivate a force which it is first of all a matter of bringing into existence: these are the apparatuses that American activists associate with the notion of *empowerment*.

Like hypnosis, like all techniques where it is a question of becoming, the techniques of *empowerment* have already been annexed by enterprises, and *empowerment* itself has already become a reference for NGOs, the commissions of UNESCO, and so on. That is why, in conclusion, I wish to link this notion to practices where it presents itself as irreducible to psychology or emancipatory good intentions: the "magic" practices of those American women (and American men) who have baptized themselves as "neo-pagan witches."

I will limit myself to an allusion, for all this is another history. If neo-pagan witches interest me, it is not only because of the test that they constitute for us who think that "we can no longer," nor because of the efficacy of the rites they elaborate and that make them capable, during demonstrations like the one in

Seattle, of bringing into existence literally "spiritual" modes of occupying public space. Nor is it because they are defined by a problematic of *healing* of which the triple—individual, collective, and ecological—dimension corresponds exactly to the triple ecological disaster diagnosed previously by Félix Guattari.[25] But it is because their risky trajectory, renewing ties with those persecuted in our past, even if it means exposing themselves to the accusation of making up a completely artificial filiation, is a product of truly pharmacological thought, discerning dangers and poisons.

> The debate about the linking of the spiritual and the political too often takes place in terms that discount or make invisible the experience of the non-dominant world. Such cultural imperialism is in itself a form of racism. It is hard for us to acknowledge that powers and dimensions of reality with which we are unfamiliar may be more than quaint hangovers from a prescientific age, that they represent people's real experience and that we might have something to learn from them. Or indeed, if we bring ourselves to admit that the dominant description of reality is too narrow, we may run slavishly after other spiritual traditions, eager to acquire experiences of non-ordinary consciousness as if they were Gucci bags or Cuisinarts, commodities that we can use to bolster our status. We become spiritual colonialists, mining the Third World for its resources of symbols and shamans, giving nothing, in a way that cheapens both the traditions we seek to understand as much as our own spiritual quests.[26]

It is certainly easier to seek initiation into a tradition that has successfully globalized itself and delegates a little bit everywhere its benevolent teachings than to do what these witches are attempting to do, that is to confront mocking skepticism without the protection of traditions that are exotic, perhaps, but culturally recognized as respectable. And the situation is even more difficult in Europe, where it is coupled with a deeply rooted contempt for the simplicity of Americans, the naivety with which, in this case, these "self-proclaimed witches" think they can "improvise a heritage," construct what has not been passed on to them. I am not a witch, I cannot testify for them, I only know that the risks they take matter to me, and that I will not be part of the sniggering. Not that I have compelling reasons I can oppose to the skeptics, but because the sniggering is here redundant in regard to the fear of ridicule that they must themselves have overcome when they learned-invented-experienced the rites able to "call forth" what, for them, would make the difference between despair or submission and what they call *healing, reclaiming, empowerment*. And that in a pragmatic mode, experiencing the effects of calling "magic" what psychology incessantly belittles, for instance under the name of suggestion. "*Witch* comes from the Anglo-Saxon root *wic*, meaning to bend or shape—to shape reality, to make magic."[27] To make magic has nothing supernatural about it, but it needs an art that they baptize with the old name borne by artisanal techniques, which demand experience and dexterity: *craft*. Outside the opposition between facts and artifacts, beyond "belief," they craft rites and learn if they transform and how. Word of immanence: "To say the Goddess is reawakening may be an act of magical creation."[28]

It is evidently possible to up the ante in such a way as to dramatize the contrast between this pragmatics, which can always be reduced to psychological terms, even to banality, a pale "as if," and the "authentic" inimitable article. We must simply call things by their name: this would be an act of war—the dirty little war that we have got so used to waging that we find it normal and legitimate to destroy what it seems possible to destroy.

The refusal to borrow from spiritual traditions coming from elsewhere, to prolong in so doing the Western gesture that defines the entire world as a sort of supermarket where one is free to acquire with impunity what is considered to be on offer, is political, but it does not take the form of a general denunciation. It engages and obliges creation. It is a "cause." The rites that witches experiment with articulate the resources and the assemblages of their world, they cross political experience and therapeutic experience according to lineages of apprenticeship that pass via feminist struggles, with the putting at risk of the private as political, via struggles having recourse to the strategy of non-violence, with the invention of techniques of consultation that produce collectives capable of such a strategy, and via learning about the poisons of mobilization, the spirit of sacrifice, and the necessity of "revolutionary discipline." And it is normal that they would be exposed to sniggers, to mocking skepticism, to the poison of irony. For it is precisely what, in them and around them, they undertake to metamorphose into power. To feel the destruction, the rape, the brutality that dismembers, like that inflicted on She who, nevertheless, returns, returns in them and through them, such that they are transformed by the force that they learn how to call forth.

> Breathe deep.
> Feel the pain
> where it lives deep in us
> for we live, still,
> in the raw wounds
> and pain is salt in us, burning.
> Flush it out.
> Let the pain become a sound,
> a living river on the breath.
> Raise your voice.
> Cry out. Scream. Wail.
> Keen and mourn
> for the dismembering of the world.[29]

# Notes

1 Léon Chertok, *L'Hypnose: théorie, pratique et technique* (Paris: Petite Bibliothèque Payot, revised and expanded edition, 1989), pp. 260–1.

2 See Léon Chertok, Isabelle Stengers, and Didier Gille, *Mémoires d'un hérétique* (Paris: La Découverte, 1990), p. 328.

3 In order to struggle against this temptation, the last chapter of *Mémoires d'un hérétique* is devoted to "complicating" (folding together) the "rapport with the world" as much as "the trance." There is no "normal" rapport with the world, but "regimes of affectation" that are distinct but capable of opening themselves up to each other. These regimes designate different ways in which a living being is affected by the world and affects it, that is to say, defines its efficacy as a "cause." Correlatively, different trances are distinguished (of repetition, vigilance, perplexity and lucidity) which at the same time are what regulate and color the different regimes and what different human techniques may stimulate by a knowingly prescribed "perturbation." "The living being, because it lives, is permanently in trance. It requisitions trances, mixes them together, intensifies them, puts them on standby, modulates them . . . That is the only adaptive strategy: the engagement with the regime of affectation that is suitable" (p. 354). Remember the "invisible gorilla": the command "pay attention"

was capable of releasing a trance of vigilance such that only "moving white T-shirts" affected us.

4   The same type of operation took place in physics and thus opposed the old physics, which "believed" in a world made of interacting particles, with "new physics," which rejoined the wisdom of the Orient. (See the writings of D. Bohm and F. Capra.) Thus, a "summit meeting" takes place between the *nec plus ultra* claimed by physics and that assigned by physicists to certain spiritual traditions that they privilege.

5   Roustang, *Qu'est-ce que l'hypnose*, p. 128.

6   Ibid.,pp. 14–15.

7   See Tobie Nathan, "The 'thing' and the 'object'," in *Nous ne sommes pas seuls au monde* (Paris: Les Empêcheurs de penser en rond, 2001).

8   Alfred North Whitehead, *Modes of Thought* (1938) (New York: The Free Press, 1968), p. 41. See also Isabelle Stengers, *Thinking with Whitehead* (Cambridge, MA: Harvard University Press, 2011).

9   Alfred North Whitehead, *Adventures of Ideas* (New York: Macmillan, 1933).

10  Deleuze and Guattari, *A Thousand Plateaus,* p. 319 (concerning terms used to describe singing birds).

11  See Bruno Latour, *On the Modern Cult of the Factish Gods*, and *Pandora's Hope. Essays on the Reality of Science Studies.*

12  Isabelle Stengers, "Le laboratoire de l'ethnopsychiatrie" preface to T. Nathan, *Nous ne sommes pas seuls au monde,* p. 41.

13  It may also be described by the effect of closing up and inhibition associated with the "black hole" in *A Thousand Plateaus*. Here "falling into a black hole" was certainly triggered by "innovative processes" then caught in a mortally stratifying system.

14  See Nathan, *Nous ne sommes pas seuls au monde.*

15  In *The Hermeneutics of the Subject,* Foucault insists on the difference between antiquity's techniques of "truth telling" and the examination of conscience, and the technique of the Christian confession. We can speak in this case of a new "machine," setting in motion Stoic assemblages and imposing on them a new regime: only the most suspicious examination of each apparently insignificant act, and the confession that it was concealing a sinful intention, can thwart the Devil's ruses (which are in the details). "Let's say, and I will stop there, that in Christian spirituality, it is the guided subject who must be present within the true discourse as the object of his own true discourse. [. . .] In Greco-Roman philosophy, rather, the person who must be present within the true discourse is the person who guides, [. . .] he is present not as the person who says: 'This is what I am', he is present in a coincidence between the subject of enunciation and the subject of his own actions. 'This truth I tell you, you see it in me.' That's it" (p. 409).

16  In this contrast, the apparatus refers to a technique, while the assemblage designates the event of the "holding together" of heterogeneous elements, conferring on each a singular mode of existence. An apparatus may be characterized in a stable manner because it is technically stabilized. On the other hand, when operations of the apparatus are described in a concrete manner, that is to say when they include the technician (always "this" technician), what is described is an assemblage (always "this" assemblage here and now).

17  The fact the therapeutic efficacy associated with taking the molecule may be superior to that of the placebo alone in no way permits the difference to be attributed to the molecule, for the idea that the effects add up is radically unfounded. It goes back to the "as if" that unites the protagonists and articulates the syntax of their pronouncements.

18  Tobie Nathan, *L'Influence qui guérit* (Paris: Éditions Odile Jacob, 1994), pp. 28–9.

19  The fact that certain of the first Freudians deemed that it would be better if artists and other creators were not analyzed is, from this point of view, very interesting.

20  See, in Nathan, *Nous ne sommes pas seuls au monde*, pp. 88–9, the importance of "this paganism—this kind of spontaneous therapeutic polytheism—of all the patients in the world, who never hesitate to bridge the supposed metaphysical oppositions between "natural" and "supernatural," between "rational" and "irrational," and address successively, sometimes even concurrently, a psychiatrist or a psychotherapist, but also a clairvoyant, a healer, a charismatic church." And do so to the displeasure of psychoanalysts who, obviously, will link the desertion of their couches to the catastrophe of an epoch which prefers expedients to the hard exigencies of the truth. See, of course, the very prescient, although astonishingly poorly written, *Pourquoi la psychanalyse?* by Elizabeth Roudinesco (Paris: Flammarion, 2001).

21  Gilles Deleuze and Félix Guattari, *What is Philosophy?* trans. Hugh Tomlinson and Graham Burchell (London: Verso, 1994), p. 108.

22  What in particular seems typical of the "ethology of sectarian beings" is the fact that if the follower wishes to be cured, to attain truth, these beings require the follower to separate from his family, to detach himself from his religion, to define the world as evil, a place of perdition (a definition confirmed by the denunciation of which the sect is a victim). *Aide psychologique aux ex-adeptes de sectes: rapport d'activité* (Centre Devereux, 2002).

23  As Tobie Nathan makes clear, the therapeutic process prescribed for ex-disciples is somewhat analogous to the one that Françoise Sironi initiated for victims of political torture—to understand the intentionality of what they were caught in. See Françoise Sironi, *Bourreaux et Victimes—psychologie de la torture* (Paris: Odile Jacob, 1999).

24  On this subject, I will refer to the passionate *The Power of their Ideas* by Deborah Meier (Boston: Beacon Press, 1995). Meier teaches in the public school system in the

East Harlem neighborhood of New York, where what is called "multiculturality" is enacted in poverty, hate and violence. The school whose thirty-year learning process she recounts has taken as a constraint not to ignore these forces but to construct a place where the teachers and their pupils learn to negotiate their power. This construction, ceaselessly taken up again, has as its axis five "habits of mind" to be adopted when confronting any question, whether it involves mathematics or a racist article in the press: "the question of facts" or "how do we know what we know?"; the question of the point of view in all these multiplicities, or "Who says that?"; the research of connections and structures or "What causes what?"; the speculation, or "How could these things have been different?"; and finally the question of knowing why that matters, or "For whom is this important?" (p. 50). No knowledge is presented as neutral, all must be approached with the same "ethological" attention, as situated, and endowed with the power to situate. A space proper to the school is thus created, a space that is always in tension and is not defined "against" these tensions, for that would separate the children from their parents, but is a space that transforms by the obligations of thought that these habits induce. And such habits enable the students to succeed in higher education, even though the milieu that they come out of would normally have excluded them from it.

25  Félix Guattari, *The Three Ecologies*, trans. Ian Sutton (London: Athlone, 2000).

26  Starhawk, *Truth or Dare, Encounters with Power, Authority and Mystery* (New York: HarperCollins Publishers, 1987), p. 19.

27  Ibid., p. 7.

28  Ibid., p. 25.

29  Ibid., pp. 30–1.

# Index

Note: Page locators followed by 'n' refer to notes.

Akhenaten 136
*Analysis Terminable and Interminable* (Freud) 101
analytic philosophers 13
anamnesis 9, 45 n.32
animal hypnosis 129
animal magnetism 2, 5, 12, 21–4, 43 n.12, 45 n.26, 59, 64 n.6, 65
animal psychologist 36
anthropomorphism 136, 137
artifact 66, 98, 110, 112, 113, 119, 121, 125
   cultural artifact 122
   facts-artifact 123, 149
   laboratory artifact 31
artisanal technique 149
autohypnosis 70–2

Bailly, Jean 25
Baruk, Stella 146
Bensaude-Vincent, Bernadette 7, 49 n.76, 50 n.80
Bergasse, Nicolas 22
Bergson 12, 46 n.42, 126 n.2
Bernheim 3, 50 n.90, 66, 67, 72, 83, 92, 96, 99
"Black Boxes" 32
Borch-Jacobsen, Mikkel 5, 6, 13, 32, 46 n.46, 50 n.85, 50 n.87, 71, 83, 97, 111, 112
bourgeois Vienna 31
Brenman, Margaret 69, 70, 78 n.6, 141

de Breteuil, Baron 21
Breton, André 3, 128 n.23
Breuer, Josef 30
Brin, David 127 n.18
Brissot, Jacques-Pierre 22
Bush, George W. 134

Carroy, Jacqueline 12
Cartesian moment 58–60
catch-all notion 87, 89, 92, 116, 133, 137, 139
Centre Georges Devereux 34, 39, 53 n.119
Chakrabarty, Dipesh 46 n.40
Charcot 2, 3, 28–30, 50 n.84, 50 n.90, 66, 67, 84, 91, 92, 99
chemistry 25, 49 n.76
Chertok, Léon 1, 4–6, 13, 14, 16, 20, 24, 26, 27, 29, 30, 39, 47 n.51, 48 n.60, 48 n.62, 49 n.74, 51 n.92, 55, 58, 84, 125 n.1, 126 n.1, 130, 131, 141
   about hypnosis 129, 130
   conferring on hypnosis 19
   *A Critique of Psychoanalytic Reason* 1, 5, 7, 17, 20, 23, 28, 33, 42 n.4
   laboratory, study of hypnosis 17–19
   *L'hypnose: Blessure narcissique* 2
   Mesmer's theory, implication of 22
   narcissistic wound 2, 56, 62
   psychoanalysis 17, 28, 30, 31
   transference neurosis 32

Christian confession 138, 151 n.15
Christian spirituality 151 n.15
Cloquet, Jules 68
cognitive psychology 83, 114
cohesive narrative 15
Cold War 3
*Consciousness Explained* (Dennett) 110
constructive philosophical
  engagement 15
contemporary hypnosis 70
contextualising hypnosis 14
continental philosopher 13
convenient arguments 12, 19
Copernican Revolution 15, 17, 19, 23, 29, 57
Copernicus, Nicolaus 56, 57, 87
*Cosmopolitics* (Stengers) 1, 8, 16, 33, 34, 41, 45 n.32, 48 n.66, 49 n.66, 49 n.76, 51 n.101, 52 n.102
countertransference 33, 35, 36, 51 n.97
Crabtree, Adam 43 n.11
*Créer le réel* (Melchior) 40
*A Critique of Psychoanalytic Reason* (Chertok) 1, 5, 7, 17, 20, 23, 28, 33, 42 n.4
cultural artifact 122
cultural imperialism 148
"The Curse of Tolerance" 33-5, 39, 52 n.105

Darnton, Robert 22
Darwin, Charles 56, 57, 87
"daughter of the Enlightenment" 8-11
deception 55, 56, 61
Delboeuf, Joseph 13, 46 n.42, 67, 76, 83, 85, 89, 92, 97, 106 n.4, 126 n.1, 126 n.2, 142
  sympathy 72
  unconscious 96
Deleuze, Gilles 16, 20, 47 n.53, 51 n.93, 51 n.101, 104, 107 n.9, 107 n.10, 107 n.11, 112, 136, 137, 144
demonic possession 5

Dennett, Daniel 110
Derrida, Jacques 13
Descartes, René 60
Deslon, Charles 23-7
Despret, Vinciane 119, 120
Devereux, George 35, 38
  ethnopsychiatry 36
  psychic reality 37
  psychoanalysis by 36
Diderot, Denis 8
disappointed 55
domestication 120, 121
double-blind medical testing 24
*Dreams of a Spirit Seer* (Kant) 12
Duyckaerts, François 46 n.42, 67

Eastern Europe 17
Eastern meditation 130
Einstein, Albert 63 n.2
*electricus* 66
*Elementary Treatise on Chemistry* (Lavoisier) 25
empowerment 41, 147
*Encyclopaedia Brittanica* 44 n.22
enigma 5
Enlightenment 8, 27
Erickson, Milton 4
Ericksonian framework 131
ethnological knowledge 62
ethnopsychiatry 34-6, 38, 52 n.102, 133, 138, 140, 147
Europe 3, 22, 48 n.62, 125, 135, 148
experimental hypnosis 2, 67, 78 n.5, 82, 110

Ferenczi, Sándor 32, 102
First World War 84
Foucauldian approach 45 n.32
Foucault, Michel 60, 62, 103, 138, 146, 151 n.15
  Christian confession techniques 138
  "modern age of the history of truth" 60, 62
  spirituality 58, 59

France   1, 2, 17, 18, 21, 52 n.102
Francophone philosophers   13
Franklin, Benjamin   22, 25
Freud, Sigmund   2, 14, 17, 28–30, 32, 36, 37, 39, 50 n.87, 56–8, 66, 67, 69, 71, 72, 74, 93, 95–100, 106 n.2, 106 n.3, 107 n.11, 137, 146
   *Analysis Terminable and Interminable*   101
   analytic technique   31
   bourgeois Vienna   31
   Copernican revolution   19
   "coup de force"   30, 100–2, 142
   *Group Psychology and the Analysis of the Ego*   73
   on hypnosis   46 n.46
   masterstroke   100, 102
   with psychoanalysis   51 n.95, 59, 118
   seduction theory   31
   unconscious   19, 93, 96, 98, 101, 141, 142
Freudian hypothesis   73
Freudian machine   9, 104, 105

Galileo, Galilei   46 n.38, 60
Gassner, Johann   64 n.6
*The Genealogy of Psychoanalysis* (Henry)   13
German idealism   12
German Romanticism   46 n.41
Gill, Merton   69, 70, 78 n.6, 141
Gille, Didier   18, 47 n.57, 47 n.58, 48 n.61
GMO   127 n.12
Gould, Stephen J.   57
Greco-Roman philosophy   151 n.15
*Group Psychology and the Analysis of the Ego* (Freud)   71, 73
Guattari, Félix   104, 107 n.10, 107 n.11, 112, 136, 148

habitual judgment   4
Hacking, Ian   110
Hall, Clark   4

hallucination   131
Hammer, A. Gordon   44 n.22
Haraway, Donna   7, 45 n.28
Harlow, Harry   119, 127 n.16, 140
healing practices   3, 9, 38–40
Hegel, Wilhelm   12
Henry, Michel   13, 47 n.47
hermeneutics   23
*The Hermeneutics of the Subject* (Foucault)   151 n.15
historical moments   3
historical research   7
history
   of human science   61
   of hypnosis   19–21, 28, 35, 39, 47 n.49, 65, 88, 92, 102, 105, 111, 134
   of magnetism   130
   of psychoanalysis   97, 101
   of science   8–10, 19, 20, 97
   "science of man"   135
   of scientific discovery   10, 11
   of scientific knowledge   56
   singularity   89
   social history   10
*A History of Chemistry* (Bensaude-Vincent)   7, 49 n.76
Hull, Clark L.   67
human science   19, 33–5, 56, 61, 63, 92
human scientific research   38
humor   8, 11, 15
hypnons   2
*Hypnoses* (Borch-Jacobsen)   5, 6, 13
hypnosis   2, 8, 12, 77, *see also individual entries*
   animal hypnosis   129
   autohypnosis   70–2
   caricatural appraisal   32
   caricature   97
   contemporary hypnosis   70
   contextualising hypnosis   14
   demonic possession   5
   efficacy   76, 78 n.3
   experimental hypnosis   2, 67, 78 n.5, 82, 110

experimental psychological
   engagements   6
heterogeneous ensemble   4
history   19–21, 28, 35, 39, 47 n.49, 65,
   88, 92, 102, 105, 111, 134
hypnotic phenomena   6, 13
hypnotic rapport   18, 29, 72, 130
hysteria   3, 5, 29, 30, 50 n.84
interest in   3
interpretations of   40, 75
knowledge practices   5, 14
laboratory-based studies   15
lucid sleep   5, 28
magic   62
by Melchior   75
mystery   62
as narcissistic wound   56, 62
new hypnosis   70, 71, 79 n.14, 115
in nineteenthcentury   61–2
objective existence   55
oblige thinking   84
obstinate emergence   13
in Paris, 1986   20
phases   131
practitioners   40
psychoanalysis (*see* psychoanalysis)
quasi-deconstructive approach   13
by Roustang   72
singularity   67
somatic account   3
theorizing hypnosis   14
therapeutic use   70
"trance" phenomena   3, 5, 61, 129
transformation of   7
true hypnosis   82
using by therapist   69
via Charcot   30
vulnerability of   85
without psychoanalysis   28
as wound   58
*Hypnosis and Suggestibility: An Experimental Approach* (Hull)   67
*Hypnosis Between Science and Magic* (Stengers)   1, 2, 33, 42

hypnotic analgesia   17
hypnotic behaviour   68
   *vs.* simulated behavior   85
hypnotic phenomena   6, 13
hypnotic rapport   15, 18, 29, 72, 130
hypnotic state   74, 123
hypnotic techniques   39, 45 n.26, 98
hypnotic tradition   126 n.2
hypnotism   13, 17, 46 n.42, 65, 77, 91, 124, 142
hypnotist   19, 44, 67, 70–6, 78 n.3, 91, 141
hysteria   3, 5, 29, 30, 50 n.84

*The Invention of Modern Science* (Stengers)   8, 20, 23, 42 n.3, 48 n.64
invisible fluid   22–4, 26

James, William   12, 13
Janet, Pierre   13
Jesuits   107 n.11
Jussieu, Antoine Laurent de   23–7, 49 n.73

Kant, Immanuel   12
Kepler, Johannes   57, 64 n.2
Keplerian ellipses   63 n.2
Koyré, Alexandre   60
Kubie, Lawrence   4, 73–5, 130
Kuhn, Thomas   127 n.11

Lacan, Jacques   14, 46 n.45, 59, 95, 102, 104, 143
Lacanian irony   60
Lacanian machine   104, 107 n.11
*La Damnation de Freud* (Nathan and Hounkpatin)   2
de Lafayette, Marquis   3
*La Nouvelle alliance* (Stengers)   16
Laplanche, Jean   46 n.46
Latour, Bruno   10, 11, 38, 52 n.102, 127 n.12, 136, 146
Laurence, Jean-Roch   43 n.12, 126 n.1

Lavoisier, Antoine-Laurent   10, 22, 24, 25, 31, 49 n.76, 52 n.103
*L'hypnose: Blessure narcissique* (Stengers and Chertok)   2
*L'hypnose entre magie et science*, see *Hypnosis Between Science and Magic* (Stengers)
liberating creativity   125
Liébault, Ambroise-Auguste   3, 50 n.90, 66
"Litvakie"   47–8 n.58
Litvaks   17
Louis XVI, King   21, 49 n.70, 88
lucid dreams   128 n.19
lucidity   19, 62, 63, 77, 78 n.13, 103, 106 n.4
lucid sleep   5, 28

McCaffrey, Anne   128 n.22
Mademoiselle Paradis   21
magic   42, 62
magic gestures   39
magnetic fluid   24–6, 65, 66, 88
magnetic lucidity   91
magnetic sleep   28, 65, 66
magnetic somnambulism   3, 43 n.13, 69
magnetism   3, 18, 92, 103, 124, 129, 130
   animal magnetism   2, 5, 12, 21–4, 43 n.12, 45 n.26, 59, 64 n.6, 65
   earlier practices   3
   history   130
   mineral magnetism   67
   momentary action   26
magnetists   3, 4
*Making Minds and Madness* (Borch-Jacobsen)   111
Mannoni, Octave   105
Méheust, Bertrand   43 n.13, 47 n.49, 50 n.85, 65, 78 n.13, 128 n.24
Meier, Deborah   152 n.24
Melchior, Thierry   40, 75

*Mémoires d'un hérétique* (Stengers)   2, 17, 19–20, 150 n.3
Mesmer, Franz Anton   3, 12, 13, 20, 23, 28, 49 n.68, 49 n.72, 62, 64 n.6, 65, 67, 88–90, 92, 104, 139
   animal magnetism   2, 22, 24
   crisis   24, 26–8, 88, 89
   hypothesized fluid   24
   invisible fluid   22–4, 26
   magnetic fluid   24–6, 66
   patient visit   21
   practice   22
   scientific ideas   23
   universal fluid   24
mesmerism   3, 8, 22, 25, 27, 49 n.70
*Mesmerism and the End of the Enlightenment* (Darnton)   22
Mesmerist movement   22
metallurgical techniques   109, 118
metallurgy   118
Michaud, Eric   13, 46 n.46
Milgram, Stanley   117, 126 n.1, 140
modern culture   5
modernist practice   29, 32, 34
modernized technique   118, 119, 122
modern knowledge practices   6, 7, 29, 33, 34
modern medicine   118
modern pharmaceutics   118, 121
modern practice   117
modern psychology   138
modern psychopathology   38
modern science   56, 58–60, 137
modern technique   118, 119
modern therapist   127 n.17
*Modes of Thought* (Whitehead)   134
Monsanto   127 n.12
multiculturality   153 n.24
music-hall hypnotizer   78 n.14

Nancy, Jean-Luc   13, 46 n.46
narcissistic wound   56, 102, 114, 137
   hypnosis as   62

Nathan, Tobie   34, 35, 37, 39–41, 48
   n.64, 48 n.65, 52 n.102, 52 n.112,
   53 n.117, 133, 138, 145, 152 n.20,
   152 n.23
   ethnopsychiatry   38
   modern psychopathology
      diagnosis   38
   non-modern healing practices
      38, 40
Nazi genocide   17
neo-pagan witches   41, 147
neuroscience   4
new hypnosis   70, 71, 79 n.14, 115
Newton, Sir Isaac   97
Newtonian force   88
Newtonian mechanics   46 n.34
Newton's gravity   22
Nietzsche, Friedrich   145
non-modern healing practices   29, 37–8,
   40, 41

objective existence   55
oblige thinking   84, 106
obstinate emergence   13
Orne, Martin   44 n.22, 68, 110, 111

Paris   2, 17, 20–2, 29, 34
perplication   34, 51 n.101
Perry, Campbell   43 n.12, 49 n.67,
   126 n.1
philosopher   5, 12, 14, 47 n.51, 59, 60,
   136, 138, 144, 145
   analytic philosophers   13
   continental philosopher   13
   Francophone philosophers   13
   knowledge practices   14
   speculative philosophers   135,
      136
philosophical engagement   8
philosophical quality   7
"Philosophy of Subjective Spirit"   12
placebo effect   113
Plato   135
Poincaré, Henri   87
*Politics of Nature* (Latour)   127 n.12

*The Power of their Ideas* (Meier)   152 n.24
pragmatic differences   6
prescientific techniques   62
Priestley, Joseph   10
Prigogine, Ilya   2, 16, 19, 47 n.55
*Principles of Psychology* (James)   13,
   46 n.43
"psyche"   4, 14, 17, 34, 58, 62, 64 n.7,
   82, 118
psychic functioning   37, 119, 124, 135,
   137, 141
psychic reality   31, 34, 37, 99–100
psychoanalysis   4, 8, 9, 13, 14, 19, 28,
   35, 36, 42 n.3, 56, 58, 62, 98,
   102, 142
   by Devereux   36, 37
   differentiate from hypnosis   14
   efficacy   103
   by Freud   59
   Freudian psychoanalysis   104
   of historians   29
   history   97, 101
   invulnerability of   107 n.11
   revisionist historians   97, 101
   rule in   69
psychoanalyst   16–18, 31, 35, 36, 62,
   71, 101
psychology   4, 18, 36, 55, 68, 83, 84, 110,
   135, 140
psychology laboratory   81
psychophysiological effect   74
psychotherapy   4, 98
psychotropic drugs   107 n.8, 122
psychotropic molecules   122
"psy" disciplines   3, 34, 43 n.13, 56
de Puységur, Marquis   3, 12, 28, 65, 66,
   69, 70, 76, 141
puzzles   127 n.11

quasi-Bernheimian view   30
quasi-deconstructive approach   13
quasi-laboratory approach   26
quasi-Lavoisierian   25
quasi-total ignorance   4
question of passivity   13

reciprocal engulfing   74
relativism   57
revisionist historians   97, 101
revisionists   106 n.2
*The Ride of Pegasus*
   (McCaffrey)   128 n.22
Roman Catholic Church   108 n.11
Rousseau, Jean-Jacques   22
Roustang, François   13, 32, 47 n.47, 48
   n.62, 48 n.65, 49 n.68, 52 n.104, 52
   n.110, 72, 131, 132
Royal Commissions, France   2

Schlegel, Friedrich   12
science   18, 22, 29, 117, 126 n.9, 138, *see
   also* human science
   behavioural science   35, 36
   development   10
   history   8-10, 19, 20, 97
   invention   9
   modern science   56, 58-60, 137
   *vs.* non-science   11
   object   92, 122
   philosophy of   12-16, 20
   role   4
"science of man"   135
scientific discovery   6, 10, 11, 97
scientific knowledge   10, 11, 29, 39, 40,
   56, 60, 86
scientific practices   8, 9, 11, 28, 32,
   122
scientific progress   7, 87, 88, 102, 106,
   107 n.8, 110, 116, 125
scientific techniques   100, 109, 117,
   118
Second World War   4, 13, 69
seduction theory   31
Shaffer, Simon   11
Shamdasani, Sonu   50 n.87
Shapin, Steven   11
shared perplexity   8
"simulation paradigm"   68
Sironi, Françoise   152 n.23
sleepwalker   2, 96
*Social Contract* (Rousseau)   22

Society for Psychical Research,
   Cambridge   3
somnambulism   76, *see also*
   magnetic somnambulism
somnambulists   28, 128 n.24
"soul"   62, 64 n.7, 84
Spinoza, Benedict   137
spiritual exercises   108 n.11
spirituality   58-60
spontaneous trances   129
Stengers, Isabelle   1, 3-5, 7, 10, 11,
   13-15, 18, 24, 26-30, 35-7,
   41, 44 n.23, 47 n.55, 47 n.57,
   48 n.61, 48 n.65, 49 n.74, 49 n.76,
   51 n.92
   about philosophy   15
   "Black Boxes"   32
   constructive philosophical
      engagement   15
   *Cosmopolitics*   1, 8, 16, 33, 34, 41,
      45 n.32, 48 n.66, 49 n.66, 49 n.76,
      51 n.101, 52 n.102
   "The Curse of Tolerance"   33-5, 39,
      52 n.105
   "daughter of the
      Enlightenment"   8-11
   history of science   8, 10
   human scientific research   38
   humor   8
   on hypnosis   13, 16, 43 n.9
   *Hypnosis Between Science and
      Magic*   1-2, 33, 42
   idea of obligation   45 n.24
   *The Invention of Modern Science*   8,
      20, 23, 42 n.3, 48 n.64
   *La Nouvelle alliance*   16
   *L'hypnose: Blessure narcissique*   2
   *Mémoires d'un hérétique*   2, 17, 19, 20,
      150 n.3
   Mesmer's theory implication   22
   modernist practice   29, 32
   non-modern practices   39
   pragmatic differences   6
   psychoanalysis (*see* psychoanalysis)
   quasi-total ignorance   4

shared perplexity   8
*Thinking with Whitehead*   1
transference neurosis   32
uniform blanket   9
*Virgin Mary and the Neutrino*   2, 15–16, 27, 49 n.76
*Studies on Hysteria* (Freud)   69
suggestibility   77, 92
   degrees of   67, 68
   mode of   67, 68
symbiotic process   16

techniques of influence   34, 35
therapeutic efficacy   32, 113, 126 n.9, 152 n.17
therapeutic intervention   62
therapeutic technique   35, 100, 127 n.12, 134, 141
therapist   40, 41, 71, 85, 99, 111, 125, 140, 141, 144
   modern therapist   127 n.17
   practicing hypnosis   95
   psychotherapists   146
   uses hypnosis   69
*Thinking with Whitehead* (Stengers)   1
*Time and Freewill* (Bergson)   126 n.2
tolerance   33, 38
traditional technique   118
"trance logic"   68
"trance" phenomena   3, 5, 61, 129, 132, 133, 150 n.3

transference   5, 30–3, 51 n.95
transference neurosis   32
true hypnosis   82

unconscious   5, 31, 52 n.111, 76, 83, 99
   Delboeuf's unconscious   96
   ego   106 n.4
   Freud's unconscious   19, 93, 96, 98, 101, 141, 142
   Lacanian unconscious   143
   personality   106 n.4
United States   3, 4, 18, 28, 43 n.11, 44 n.17, 110
universal fluid   24

*Virgin Mary and the Neutrino* (Stengers)   2, 15–16, 27, 49 n.76
von Lieben, Anna   30

Western colonialism   9
western culture   5
Western society   140
Whitehead, Alfred North   1, 2, 47 n.52, 50 n.83, 134, 135
Wittgenstein, Ludwig   13
wound   57, 58, 61, 63 n.2, *see also* narcissistic wound

Yiddishland   17

Zen meditation   129